Innovative Tableau
100 More Tips, Tutorials, and Strategies

Ryan Sleeper

Beijing · Boston · Farnham · Sebastopol · Tokyo

Innovative Tableau

by Ryan Sleeper

Copyright © 2020 Playfair Data. All rights reserved.

Printed in Canada.

Published by O'Reilly Media, Inc., 1005 Gravenstein Highway North, Sebastopol, CA 95472.

O'Reilly books may be purchased for educational, business, or sales promotional use. Online editions are also available for most titles (*http://oreilly.com*). For more information, contact our corporate/institutional sales department: 800-998-9938 or *corporate@oreilly.com*.

Acquisitions Editor: Jonathan Hassell	**Indexer:** Judith McConville
Development Editor: Angela Rufino	**Interior Designer:** David Futato
Production Editor: Katherine Tozer	**Cover Designer:** Karen Montgomery
Copyeditor: Octal Publishing, LLC	**Illustrator:** Rebecca Demarest
Proofreader: Piper Editorial	

May 2020:   First Edition

Revision History for the First Edition

2020-04-24:   First Release
2021-05-14:   Second Release

See *http://oreilly.com/catalog/errata.csp?isbn=9781492075653* for release details.

The O'Reilly logo is a registered trademark of O'Reilly Media, Inc. *Innovative Tableau*, the cover image, and related trade dress are trademarks of O'Reilly Media, Inc.

978-1-492-07565-3

[MBP]

Table of Contents

Part II.   More Chart Types

Part III.   Author Experience

Part IV.   User Experience

Preface

Data is everywhere today. Technological advances have made it possible to track and store more data than ever before, providing a competitive advantage for those who can best harness and translate this twenty-first-century "oil." The best way to convert raw data into valuable insights is through analytics and data visualization, and—just like we need refineries to transform oil into more useful products—we need tools to help us capture the potential competitive advantage data can provide.

In a crowded field of analytics and data visualization tools, Tableau's pace of innovation has helped it emerge as a clear leader in this space for the foreseeable future. As of this writing, Tableau has been recognized as a leader in Gartner's Magic Quadrant for Business Intelligence and Analytics Platforms for eight consecutive years. In 2019, Tableau was acquired by Salesforce, which will only further expedite Tableau's trajectory and make its products an everyday need for hundreds of thousands of users around the world. Simply put, there has never been a better time to learn and level up in Tableau.

One secret to Tableau's success is that not only is it consistently providing product innovations, but the company also provides a platform with a unique flexibility in the market which allows its users to create their own innovations. You can now design dashboards down to the pixel, integrate third-party applications through dashboard extensions, refine approaches within the product, and even invent new chart types.

This Book's Purpose

I wrote *Innovative Tableau* as a sequel to *Practical Tableau: 100 Tips, Tutorials, and Strategies from a Tableau Zen Master* to build on the foundation laid out in that book and share innovations that will help us collectively change the world with data.

This book has three objectives:

1) To catch you up on the product developments introduced since the release of Practical Tableau

> Although Tableau is almost constantly introducing new product features, it also does an uncanny job at keeping the heart of the product intact. In fact, of the *100* chapters in *Practical Tableau*, only a single chapter has become obsolete since it

was first released in 2018—and that's due to a change in the company's license model; not the product itself.

For this reason, we chose to publish a completely new sequel as opposed to a book update. However, we want to bridge the gap between the first and second titles by filling you in on what's new and keeping you on the cutting edge of what Tableau is capable of.

2) To help you level up

To me, the biggest compliment is when somebody asks, "You made that in Tableau?" This is not to say that Tableau doesn't help you make beautiful visualizations and seamless user experiences on its own, but it means that you have an opportunity to set your work apart. This book is a guide for getting the most out of Tableau and pushing the envelope on what you can deliver.

3) To inspire creative applications of the fundamentals

You will come to realize that you can combine individual techniques to create what I call "creative applications of the fundamentals." This book shares many of my favorites in the hope that you can immediately improve your own and your audience's experience with Tableau. Most important, I hope it inspires you to think differently when you're using Tableau and sparks your own creations that you will pay forward.

This Book's Audience

This book is relevant for anybody who wants to get more from Tableau, make more engaging and effective chart types, improve their own authoring experience, and deliver world-class analytics tools to their audiences.

That being said, it is assumed you have read the first book in this series, *Practical Tableau*, and/or have the prerequisite level of Tableau experience to match the content in the first book.

This Book's Structure

The one hundred chapters in this book are organized in four parts:

Part I

You'll get my updated advice on how to learn Tableau and be introduced to my favorite product capabilities that have been released since *Practical Tableau* was published.

Part II

Practical Tableau included *29* tutorials on making a variety of charts, and this book includes *36* more! You'll go beyond the Show Me button and read tips on making fundamental chart types more engaging, integrating technical features to make charts more effective, and creating nonstandard chart types and designs, including many you won't find anywhere else.

Part III

You'll see some of my favorite tutorials for making your life easier as a visualization author.

Part IV

This final part includes some of my favorite tutorials for making your audience's life easier as an analyst.

Many tutorials in this book were created using Tableau Public and can be downloaded from my Tableau Public portfolio (*https://oreil.ly/XMjQt*).

Conventions Used in This Book

The following typographical conventions are used in this book:

Italic

Indicates new terms, URLs, email addresses, filenames, and file extensions.

`Constant width`

Used for program listings, as well as within paragraphs to refer to program elements such as variable or function names, databases, data types, environment variables, statements, and keywords.

`Constant width bold`

Shows commands or other text that should be typed literally by the user.

This element signifies related content from *Practical Tableau* (O'Reilly, 2018).

This element signifies a general note.

O'Reilly Online Learning

 For more than 40 years, *O'Reilly Media* has provided technology and business training, knowledge, and insight to help companies succeed.

Our unique network of experts and innovators share their knowledge and expertise through books, articles, and our online learning platform. O'Reilly's online learning platform gives you on-demand access to live training courses, in-depth learning paths, interactive coding environments, and a vast collection of text and video from O'Reilly and 200+ other publishers. For more information, visit *http://oreilly.com*.

How to Contact Us

Please address comments and questions concerning this book to the publisher:

> O'Reilly Media, Inc.
> 1005 Gravenstein Highway North
> Sebastopol, CA 95472
> 800-998-9938 (in the United States or Canada)
> 707-829-0515 (international or local)
> 707-829-0104 (fax)

We have a web page for this book, where we list errata, examples, and any additional information. You can access this page at *https://oreil.ly/innovative-tableau*.

Email *bookquestions@oreilly.com* to comment or ask technical questions about this book.

For news and information about our books and courses, visit *http://www.oreilly.com*.

Find us on Facebook: *http://facebook.com/oreilly*

Follow us on Twitter: *http://twitter.com/oreillymedia*

Watch us on YouTube: *http://www.youtube.com/oreillymedia*

Acknowledgments

I owe a tremendous debt of gratitude to Tableau's team of developers for creating a platform that allows so many of us to make a positive impact with data—and have fun doing it. Most of you do not get much public glory, but I hope you are well-aware that your contributions are immeasurably grand on a global scale.

There are also the present and past Tableau employees, many of whom have helped shape my career. Thank you for your mentorship and friendship, Alivia Hale, Amanda Boyle, Andy Cotgreave, Ben Jones, Brad Welch, Colby Pash, Daniel Hom, Dash Davidson, Edward Beaurain, Elissa Fink, Ellie Fields, Jennifer Nguyen, Jewel Loree, John Iwanski, John Jensen, John Wilson, Jonah Kim, Jordon Scott, Kevin Krizek, Kyle Gupton, Lauren Rogers, Mac Bryla, Michael Long, Ross Perez, Amy Sarah Elliott, Scott Teal, Sophie Sparkes, and Tracy Rodgers.

The most unique thing about Tableau is its community. It is difficult to describe if you have not experienced it, but the best way I can summarize is to say Tableau has somehow managed to create a community of people, based on a software, with whom I genuinely want to hang out. Just good folks trying to help one another improve our careers and the world around us.

I never thought it would be possible to make so many friends through a software program I use for work. These people have played a critical role in pushing one another to new heights, motivating me personally, and just making life more fun. This is where I am going to fail miserably at recognizing everyone who has inspired me along the way. Thank you Adam Crahen, Adam McCann, Andi Haiduk, Andrew Kim, Andy Kriebel, Ann Jackson, Anthony Armstrong, Anthony Chamberas, Anya A'Hearn, Ben Sullins, Brandi Beals, Bridget Cogley, Brittany Fong, Carl Allchin, Cesar Picco, Chloe Tseng, Chris Love, Christopher DeMartini, Christopher Scott, Corey Jones, Craig Bloodworth, Curtis Harris, Dan Murray, David Murphy, David Pires, Emily Kund, Emma Whyte, Jacob Olsufka, Jason Harmer, Jason Penrod, Jeff Plattner, Jeffrey Shaffer, Jeremy Poole, Jim Donahue, Jim Wahl, Jonathan Drummey, Josh Tapley, Josh Jackson, Joshua Milligan, Justin Hinckfoot, Keith Helfrich, Kelly Martin, Ken Flerlage, Kevin Flerlage, Lindsey Poulter, Lorna Eden, Lucas Brito, Mark Jackson, Matt Chambers, Matt Francis, Matt Hoover, Michael Perillo, Mike Cisneros, Mike Moore, Neil Richards, Nelson Davis, Pablo Gomez, Patrick McCormick, Patrick Moore, Paul Banoub, Paul Chapman, Peter Gilks, Pooja Gandhi, Ramon Martinez, Rina Petersen, Rob Radburn, Rody Zakovich, Sarah Bartlett, Sarah Nell-Rodriquez, Sean Miller, Shine Pulikathara, Simon Beaumont, Skyler Johnson, Steve Bennett, Steve Fenn, Steve Wexler, Ty Fowler, Will Perkins, Will Strouse, Yamil Medina, and Yvan Fornes.

And, most important, thank you to my wife, Amy. This book, business, and life, would not be possible without you.

PART I

Introduction/What's New

CHAPTER 1

How to Learn Tableau

The single most-common question I'm asked is, "What advice do you have for someone getting started with Tableau?" In *Practical Tableau*, I covered my top five tips for learning Tableau, but as the product itself evolves, so do the resources for helping you get the most out of the software. This chapter provides my top tips for getting the most out of this book and beyond as you continue your Tableau journey.

Read

Look at you—already off to a great start! I am naturally partial to the book that you are reading now, but there are several great authors who have inspired me, and Tableau books from which I have learned.

For visualization applications beyond both *Practical Tableau* and *Innovative Tableau*, I recommend *Communicating Data with Tableau* (O'Reilly, 2014) by Ben Jones, founder of Data Literacy.

For additional technical know-how, I recommend reading anything from Joshua Milligan, including *Learning Tableau* (Packt, 2019).

For inspiration, I recommend *The Big Book of Dashboards* (Wiley, 2017) from Andy Cotgreave, Jeffrey Shaffer, and Steve Wexler. This is not a Tableau-specific book, but many examples were built in Tableau, and it provides a chance to review critiques from some of the best in the industry.

As for *Innovative Tableau*, keep an eye out for the kingfisher icon throughout the rest of this title. This reader aid indicates when a prerequisite for the chapter you are reading has been covered in greater detail within *Practical Tableau*.

Introducing Playfair Data TV

Since the release of *Practical Tableau*, my company has launched its own on-demand Tableau training platform, Playfair Data TV (*https://oreil.ly/P_KMI*). With more than one hundred video tutorials and new content released every month, Playfair Data TV is already one of the largest and fast-growing collections of Tableau video tutorials available. This type of video training helps complement and reinforce your learning with additional context.

Other video instructors and platforms I recommend are Matt Francis's Udemy course and the content from Pluralsight's all-star team of authors, including Adam Crahen, Pooja Gandhi, Ann Jackson, Lilach Manheim, Curtis Harris, and Michael Mixon. Each of these authors selflessly gives their knowledge away, and I am grateful for their contributions.

Playfair Data TV was launched to complement the expanding video training resources available with some new perspectives on taking a strategic approach to Tableau projects, applying storytelling techniques to Tableau visualizations, and using Tableau tactics in real-world applications. You will find that the format of my video content mirrors that of *Innovative Tableau,* and you can find the video versions of several chapters there.

Jump into the Community

I mentioned in the Preface that the Tableau community is difficult to explain because it really is so unusual in the world of software. Going to their conference is closer to a family reunion than a trip for work. One of the best things about it is that I can personally guarantee you that the Tableau community will embrace you if you want to join. Here are just a few ideas on how the Tableau community can help your learning progress:

Join a local Tableau User Group
> Tableau has established user groups all over the world that typically hold monthly or quarterly meetings. These are groups that are run by and for the community where speakers provide tutorials and/or use cases that are likely relevant for your own work. These meetings are also a great place to network and discover how others are using Tableau to solve their unique business problems.

Get social
> The Tableau community is full of avid social media users, particularly on Twitter. What I like most about this channel is that it provides a source of inspiration, encouragement, and built-in peer review should you choose to share your work.

Use the forums
> I still don't know how they do it, but rarely does a valid forum question go unanswered. If you find yourself stuck and have already checked to ensure a similar use case has not been solved, try to post your question on Tableau's user forums. You will likely receive help, and your question and its solution will help somebody searching in the future.

Attend a Live Training

If the books, videos, and community just can't quite make things click for you—or if you just want a jumpstart to mastering Tableau—an in-person training may be for you. As much as I try to be thorough and structure my writing in an intuitive way, sometimes you just need a live trainer to help you connect the dots. For this reason, we have developed our own series of Tableau training workshops (*https://oreil.ly/ hF708*) covering Fundamentals, Advanced, and Tableau Prep curriculums, which are hosted throughout North America.

These events are more like mini conferences than technical trainings, with networking breakfasts and lunches, one-on-one coaching, inspiration, and lots of giveaways!

I also vouch for Tableau's own two-day trainings, which you can find just about anywhere. My career benefited greatly in my second year using Tableau from attending an advanced training taught by Molly Monsey.

Practice with Tableau Public

As far as I'm concerned, Tableau Public is still the undisputed best way to improve with Tableau. Tableau Public is a free version of Tableau that has almost all of the same capabilities as Tableau Desktop. The catch is you must save your work to the public web where anybody can see it, so it is not a suitable solution for private work data. I view this as a benefit because it forces you to use data outside of work.

Tableau Public has a way of becoming a sandbox in which you can experiment with data you are personally interested in; inevitably, you will pick up new techniques that you can then apply at your day job. Not to mention that Tableau Public provides a portfolio of hundreds of thousands of visualizations from which you can draw inspiration. You can download many of these workbooks and then reverse-engineer them and pick up technical know-how.

I always say that if you buy into my training ecosystem, including the book you are reading now, I can help get you to roughly the eightieth percentile of Tableau users worldwide—but the rest of the way is up to you. Learning Tableau is like learning a foreign language, and no matter how much you learn, if you do not practice, your skills will deteriorate over time. Tableau Public is the best tool for staying sharp and perfecting the skills that you'll acquire in the coming chapters.

CHAPTER 2

Parameter Actions

The chapters in Part I don't necessarily go in order of importance or chronology, but I'm declaring author's prerogative and starting off with my personal favorite: *parameter actions*. I have always described Tableau parameters as a Swiss Army knife because of their infinite applications. In case you are not familiar, Tableau parameters are user-generated values that allow you and your audience to choose the inputs of calculated fields. You, as the author, code the allowable values once, but then it's up to the user as to which value is selected.

This unlocks a higher level of flexibility in Tableau that allows users to manipulate analyses on the fly by using a *parameter control* to change the values populating calculated fields. As of Tableau Desktop version 2019.2, there is a Change Parameter action, which makes the ability to change parameter values even more seamless.

How to Use Tableau Parameter Actions

Parameters are global values (i.e., not attached to a specific dataset) that can be controlled by a workbook user. These values can be used within calculated fields, filters, histogram bin sizes, and reference lines. You, as the dashboard developer, get to put some *parameters* around what those values can be, but from then on the end user gets to decide which value populates a calculated field or filter.

 Related: *Practical Tableau*, Chapter 14, "An Introduction to Parameters" (O'Reilly, 2018)

For an introductory example, we use a sales-by-month trend in the Sample – Superstore dataset to show you how clicking it will overwrite the current value of a parameter.

To create a parameter, I like to right-click any blank space in the Data pane and then click Create Parameter. You can also click the down arrow in the upper-right corner of the Dimensions area of the Data pane and then click Create Parameter. Here's how it looks if I were to make a new parameter in the Sample – Superstore dataset:

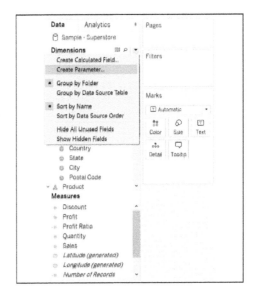

To illustrate how to use the newest type of action, here's a parameter with a data type of Float and allowable values of All:

These settings mean that this parameter can be populated with *any* number including those with decimals. Next, I set up a simple sheet with a mark type of Text to show you the current value of the new parameter:

Note that I also changed the format of the number being displayed to Currency (Standard), which for me (in the US) has a dollar sign and two decimal places to match the format of the numbers in the next step. Because the default, current value of the Parameter Actions Example parameter is 1, we see a result of $1.00 displayed.

Next, I set up a line graph that looks at the measure of Sales in the Sample – Superstore dataset by continuous Month of Order Date and place both sheets onto a dashboard together:

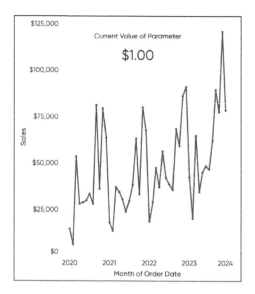

To add a parameter action from the menu bar at the top of the window, click Dashboard and then Actions. Click the Add Action button; you will see six options including the new Change Parameter action:

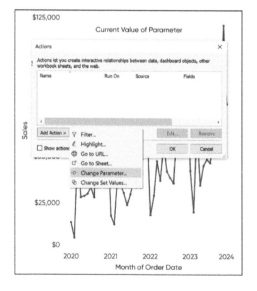

In the dialog box that opens after you click Change Parameter, you can control the settings such as the source sheet of the action, when you want it to run, which parameter is overwritten, and based on which value. In this example, let's specify that clicking the line graph will overwrite the Sales value in the Parameter Actions Example parameter:

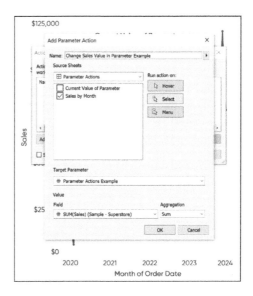

After you click OK to save the settings, clicking any Sales value in the line graph changes the value being displayed on the Current Value of Parameter sheet:

Because we set up a Change Parameter action, clicking a value within the line graph is overwriting the current value of the Parameter Actions Example parameter, which is feeding the Current Value of Parameter text sheet.

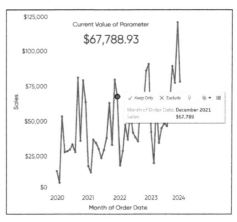

At this time, you do not have the option to set up what to do when you clear a parameter action, but when the data type is Float like this first example, you *can* do a multiselect. Because the aggregation of our parameter action is SUM, clicking more than one data point will add them together:

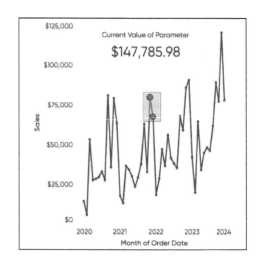

For this introductory example, we changed the parameter value for a parameter with a data type of Float, but you can also set these up to work with integers, dates (and times), strings, and Boolean values.

CHAPTER 3

Set Actions

As of Tableau Desktop 2018.3, you can also use actions to change set values. I started with parameter actions because parameters are the key to many of my tips and tricks within Tableau. Parameters provide complete control over the user-generated inputs being selected (i.e., the allowable values), but set actions provide their own unique flexibility. Most notably, set actions allow you to do the following:

- Multiselect, even when the data type is String (parameters require a single select with qualitative data types)
- Use empty sets (parameters require a value to be selected)
- Instruct Tableau as to what to do when the selection is cleared (parameter actions do not currently provide this option).

How to Use Tableau Set Actions

To illustrate how to use Change Set Values actions, let's design a user experience that allows users to control the values in a grouped bar chart by clicking custom regions on a map in the following dashboard:

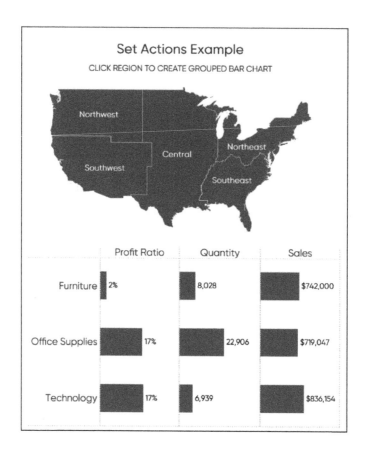

Here is how the Map view started out in the Authoring interface:

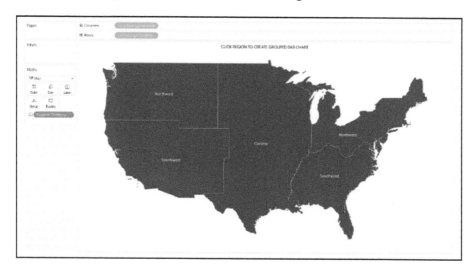

Note that this map is created from a dimension called Custom Territory to draw the region borders and labels. You can learn how to make maps with custom territories in Chapter 29.

To use set actions, there must be a set on the view. Because I want to allow my user to click one or more regions to control the bar chart, I create a set out of the Custom Territory dimension. Here's the easiest way to create a set: within the Authoring interface, right-click the dimension from which you want to create the set, hover over Create, and then, on the menu that opens, choose Set:

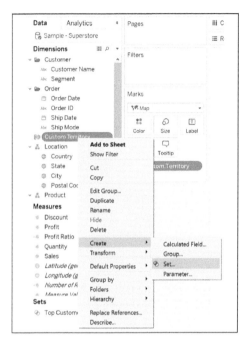

This pulls up the settings for the set where you can give the set a name and control which dimension members are included in the set. Here, I rename the set but leave it empty for now because we will control the values included in this set with actions in a future step:

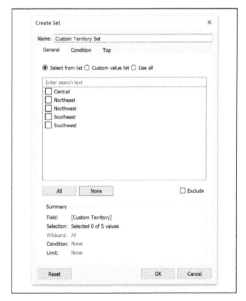

Again, to use set actions, the set must be on the view somewhere. In this case, I color the custom regions by whether they are in the set by placing the newly created Custom Territory Set on the Color Marks Card:

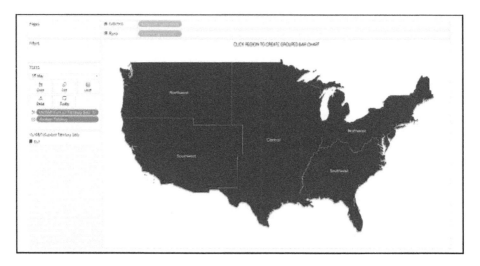

This colors the dimension members in the set one color, and everything else a second color. We see only one color so far because there aren't any dimension members in the set.

In this use case, we are controlling the bar chart, so we also need this newly created Custom Territory set somewhere on the Grouped Bar Chart view. Here is where the bar chart view started:

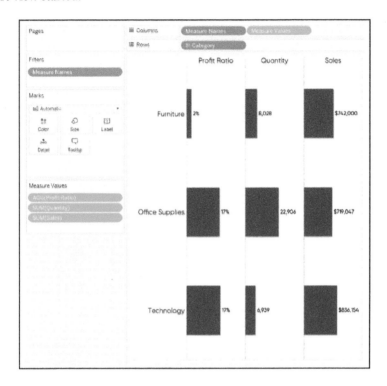

To get the same Custom Territory Set being used on the map view somewhere on this view, I convert the bar chart to a grouped bar chart by placing the set on the Rows Shelf after the Category dimension. This results in two subrows for each Category: one for the measure values *in* the set; one for the measure values *out* of the set. For extra emphasis, I also add the Custom Territory Set to the Color Marks Card. Lastly, I hide the In/Out Header by right-clicking the pill for the set on the Rows Shelf and deselecting Show Header:

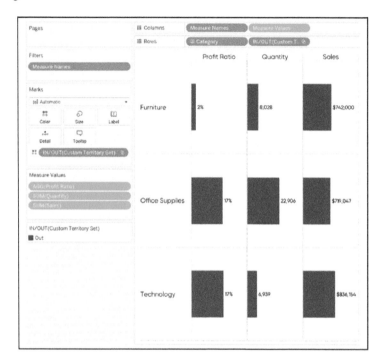

We don't see anything change on this chart yet, because there are no dimension members currently in the set. We are going to change that after adding both sheets to a dashboard and adding a dashboard action. To create a set action, on the menu bar of the Dashboard interface, click Dashboard, and then choose Actions:

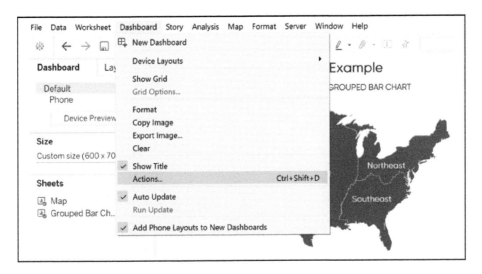

In the dialog box that opens, click the Add Action button and then, on the drop-down menu, choose Change Set Values:

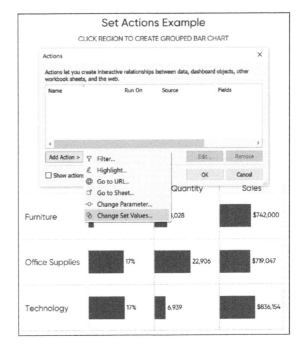

From here, we can control which sheets the action originates, when we want the action to run, what we want to happen when we clear the selection, and, most important, which set's values we want to overwrite with the dimension members selected in the source sheet(s).

Here are the settings for my use case, which are instructing Tableau that if dimension members are clicked on the Map view, overwrite the dimension members in the Custom Territory Set. Then, if you clear the selection, which you can do by clicking the Escape key, all values will be removed from the set and the view will reset as if nothing were clicked:

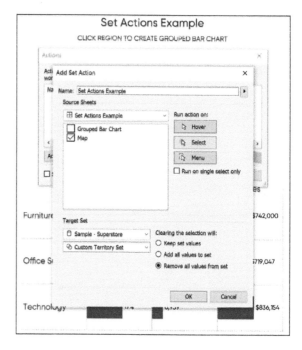

After clicking the OK buttons to close the settings for my action, clicking one or more custom territories on the map will overwrite the dimension members included in the Custom Territory Set, and because we're coloring both charts and controlling the rows in the grouped bar chart by that set, we will see the dashboard update. Here's how it looks after clicking the Central territory:

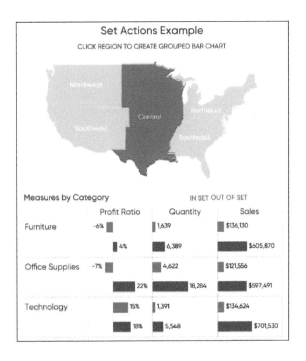

Note that I also added a title to the Grouped Bar Chart sheet, which doubles as a color legend.

If I were to clear the selection by pressing the Escape key (or by clicking the Central territory again), the dashboard reverts as if nothing happened:

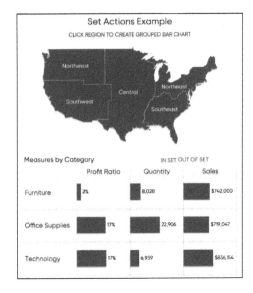

And because sets can contain none, one, or *more* dimension members, I can also do a multiselect. Here's how the view looks if I click the Northwest custom territory and then press and hold the Control key while I click the Southeast custom territory.

One notable insight is that this set containing the Northwest and Southeast regions has a much better profit ratio in the Furniture category compared to the combination of the Southwest, Central, and Northeast territories, but much lower sales. Perhaps we can learn what these territories are doing to maximize their profit ratio so that we can use similar tactics in other territories or figure out how to raise sales in these highly profitable areas.

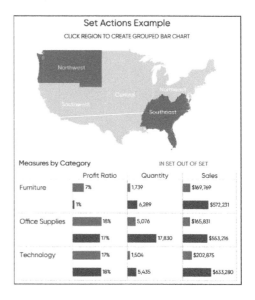

Actionable insights made possible through set actions!

CHAPTER 4

Button Dashboard Object

In Tableau Desktop 2018.3, Tableau gave us the long-overdue Button dashboard object! Before this new object was released, adding buttons to a dashboard required a clever combination of worksheets and dashboard actions (*https://oreil.ly/sJQcm*). I say this feature was long overdue because before it came out, this was my second-most-popular blog post, *ever*.

When a hack is that popular, it is time to make it a standard feature in the product. Fortunately for us, one source of Tableau's innovation is truly its user base. Not only do the developers listen to the community, they even have an Ideas space on Tableau community forums where user ideas are collected, voted on, and updated as each successful idea's status transitions from submittal to production.

This chapter shows you how to use one of Tableau's newest dashboard objects to link to dashboards or worksheets within the same workbook.

How to Use the Button Dashboard Object in Tableau

As I mentioned in Chapter 1, I credit much of my success with Tableau to its free tool, Tableau Public. Tableau Public provides a source of inspiration, a platform for creating a data visualization portfolio, a sandbox for trying new ideas, and a treasure trove of tactics to reverse engineer that so that you can incorporate approaches into your own work.

For these reasons, several of the tutorials in *Innovative Tableau* are illustrated using the following Tableau Public workbook that I affectionately call *Super Sample Superstore*:

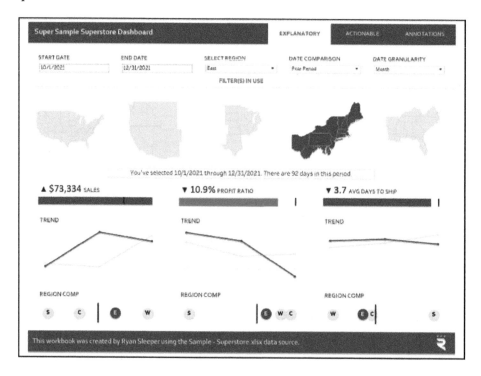

You can download the workbook if you want to follow along (*https://oreil.ly/_gPPm*) and then, in the lower-right corner of the window, click the Download button.

The workbook contains three dashboards: Explanatory for a high-level overview, Actionable to reveal insights that can cause action, and Annotations to add custom commentary.

For this example, let's add a button to the Actionable dashboard:

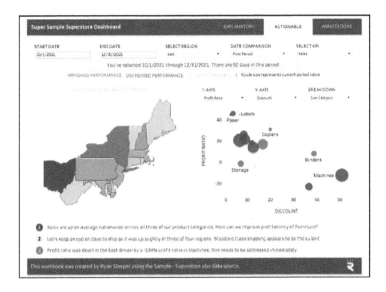

At the bottom of the dashboard, you can see custom insights and recommendations that have been added by an analyst (see Chapter 84). You also see a link to add those annotations in the top navigation (see Chapter 72), but because our user likely hasn't previously seen this option, let's add an extra button to guide them to the Annotations dashboard.

To add a button that links to another sheet within the same workbook, go to the Dashboard pane on the left side of the main Dashboard interface, and then, from the Objects area, simply drag a Button object onto a dashboard:

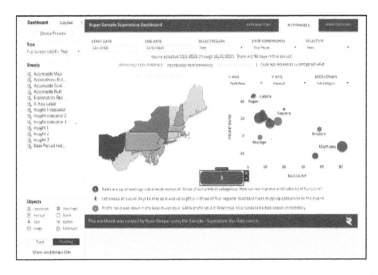

Starting in Tableau version 2020.1, the Button object was renamed *Navigation*.

In this example, I have added a floating Button object to the Actionable dashboard within the *Super Sample Superstore* workbook. By default, you will see a big gray button appear. These buttons can be made of images or text. You can also choose which sheet a button links to, change the formatting options, and even inform the user what the button does via its tooltip.

To access all of these options, select the Button object, click the down arrow that appears on the right of the object and then, from the drop-down menu that opens, choose Edit Button:

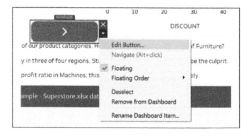

The Edit Button dialog box opens in which you can configure the navigation and formatting for the button.

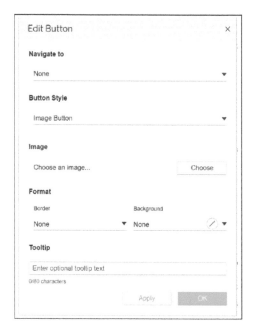

For this example, I set up the button to appear as an image of a speech bubble, link to the Annotations dashboard, and provide a call to action in the tooltip:

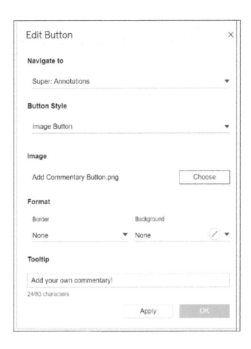

After you click OK to accept these settings, you can switch to the Layout pane where you can control the button's size and location on the dashboard (assuming that the Button object is floating):

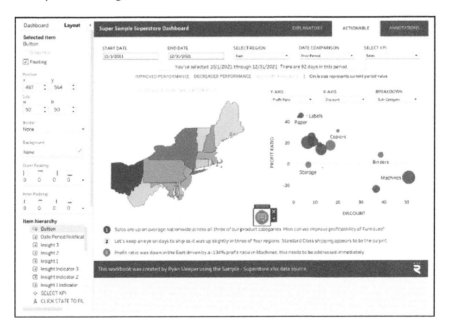

Now, when a user hovers over the button, the call to action displays, encouraging them to add their own commentary. If they click the button, it brings them to the Annotations dashboard within the workbook! You can test out the user experience of a button by going into Presentation mode within Tableau Desktop (by clicking F7).

The Button dashboard object is a welcome update for linking within the same workbook. However, if you need to link *outside* of the workbook, your best bet is to add an Image object, instead, and then set the object's Target URL.

CHAPTER 5

Transparent Sheets

As of Tableau Desktop 2018.3, you can make the backgrounds of dashboard worksheets transparent. You have been able to use a combination of floating and tiled objects since all the way back to Tableau version 8, which allows you to create what I call *interactive infographics* by floating interactive elements over a static background image.

The ability to make the backgrounds of floating objects transparent provides a flexibility that leads to significantly more possibilities. Just to name a few, you can do the following:

- Improve the design of interactive infographics by integrating the background image with the sheet in the foreground
- Place charts in better spatial context of each other
- Go beyond a dual axis by layering additional worksheets on top of each other

This chapter shows you how to make the backgrounds of worksheets transparent in Tableau.

How to Make Worksheet Backgrounds Transparent

For this example, suppose that we want to make the following bar chart more engaging by adding a custom image behind it:

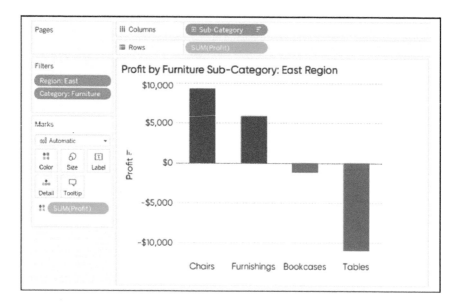

First, let's start a new dashboard and add a tiled Image object so that it fills the entire dashboard:

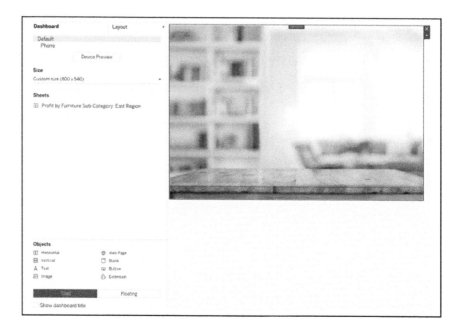

Next, we place the worksheet containing my bar chart on the dashboard as a floating object:

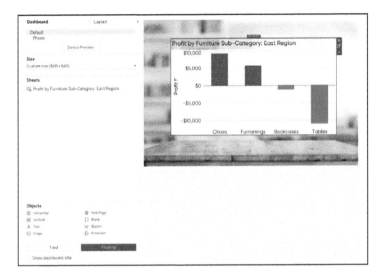

This infographic is *OK*. We have a nice professional background, and the picture of the table at the bottom is clear to really drive the focus of the visualization home. But note that by default the *title* of the worksheet is transparent, which makes it much more integrated with the background image. I'd like to apply this same effect to the rest of the worksheet, which, fortunately, we're now able to do.

To do this, right-click anywhere within the worksheet, choose Format, and then, on the Format pane, click the paint bucket icon (i.e., Shading tab):

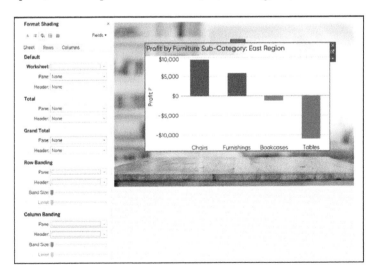

By default, worksheets have a white, opaque background. To change this to transparent, click the Worksheet drop-down palette and select None. The white background goes away and the chart is now fully integrated with the background image:

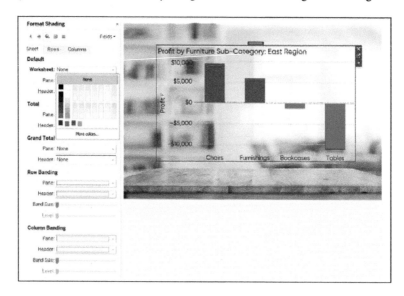

To finalize the view, let's make the font bold and add a floating Text object that acts as a dashboard title:

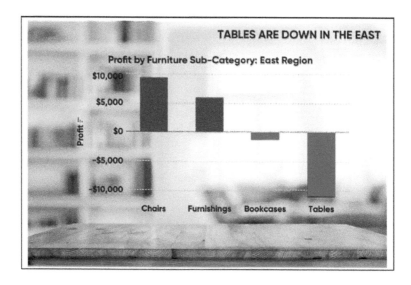

How to Make Map Backgrounds Transparent

Maps are slightly different than other worksheets because they are made up of a series of layers. You can make them transparent just as I've described so far, but there are a couple of extra steps to take. As with our first example, right-click anywhere on a map, choose Format, and then change the Worksheet shading to None:

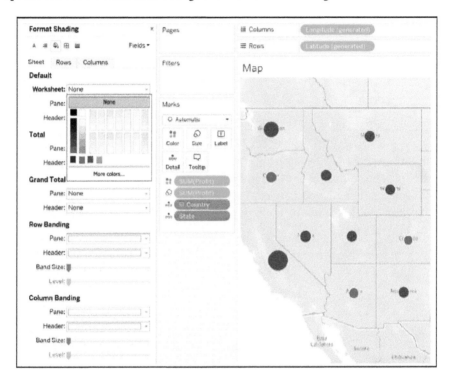

With maps, the Base layer is opaque, so you must turn off that layer. You can toggle Map layers on and off by clicking Map in the top menu bar and then choosing Map Layers. Then, in the Map Layers area, clear the checkbox to the left of Base:

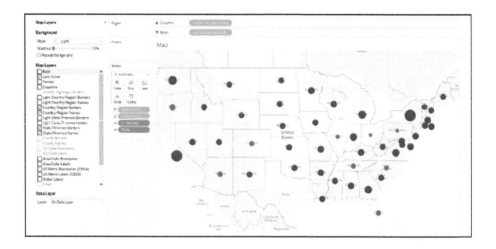

Turning off the Base layer gives you a map that has a completely transparent background that you can float over other images, charts, or maps. Note that the Base layer provides some differentiation between land and water, so you will likely want to select the Coastline checkbox to turn on that layer and bring back the appearance of borders:

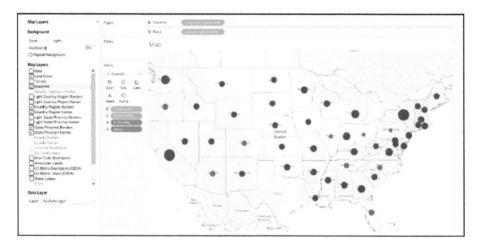

I would be remiss in both of these examples if I didn't point out that when you integrate background images like this, there is a fine line between making an insight engaging and *chartjunk* (which we discuss more in Chapter 12). This technique works best when the background image has large areas of like or faded colors that don't interfere with the visualizations layered on top of it.

CHAPTER 6

Show/Hide Button on Layout Containers

If making the backgrounds of sheets transparent were not enough, as of Tableau 2019.2, you can toggle floating dashboard layout containers on and off completely. This means that if you place a chart within a layout container, you can choose to show or hide that entire chart with the click of a button!

The ability to toggle layers on and off lends itself to some improved user experiences because you can show or hide elements such as instructions, filters and parameter controls, or additional layers of context. The ability to choose when to show or hide containers even provides a more innovative way to do *sheet swapping*, as described in Chapter 46.

This chapter shows you how to add a Show/Hide button to a layout container in Tableau.

How to Toggle Layout Containers On and Off

For this example, let's revisit the Actionable dashboard within the *Super Sample Superstore* workbook (*https://oreil.ly/5K2OG*) from Chapter 4. Here's our scenario: because this is the second dashboard in the workbook, and it is assumed that most of the controls will be adjusted on the first dashboard (Explanatory), we want to make that row of filters and parameter controls less of a focus.

Here's how the original dashboard looks:

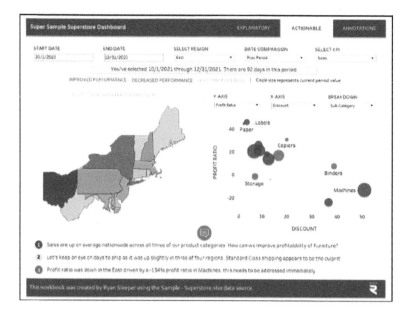

And here's an updated version after removing the controls and moving the notifications up to provide better spacing for the visualizations:

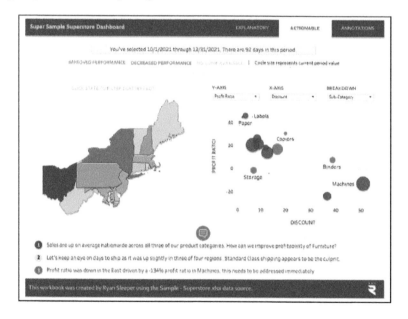

This version is slightly cleaner and we might now even have room to add a fourth insight at the bottom, but we have lost the ability to change the workbook's settings without returning to the first page. To provide the best of both worlds—clean dashboards and flexible user control—let's set up a container with the controls that we can toggle on and off!

First, in the Objects area of the Dashboard pane, add a floating Horizontal layout container to the dashboard. This works with both Horizontal and Vertical layout container objects, but here we use a Horizontal container because the controls are in a horizontal orientation:

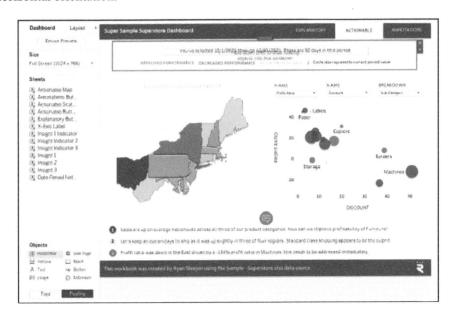

Note that I have placed the container over the date period notification and color legends, which is not ideal temporarily, but we will eventually turn this container off.

It's also good practice with this use case to make the background of the layout container opaque. To do this, select the container object, click to open the Layout pane, and then, in the Background pane, select a color that matches the rest of the dashboard:

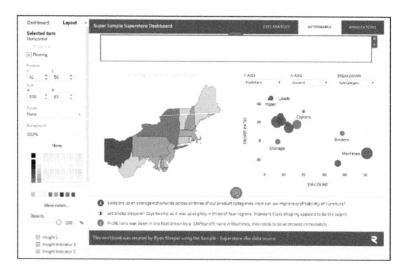

Next, place all of the elements that you want toggled on and off within the container. My controls include a series of filters and parameter controls that I previously removed from the dashboard. To get them back, simply select any sheet and then, in the upper-right corner of the sheet, click the down arrow to open a drop-down menu, hover over Filters, Parameters, or whatever else you want to add, and then make your selection:

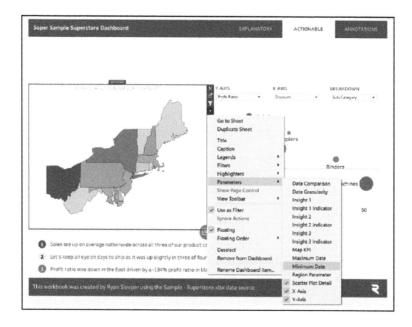

Here's how my dashboard looks after adding my controls to the layout container:

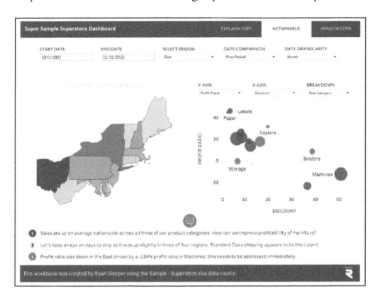

Now we're ready to add the new Show/Hide button. To do so, select the layout container. The easiest way to select a container is to click any object within it and then double-click the gray rectangle that appears at the top of the selected object. A blue border around the container indicates when you have selected it:

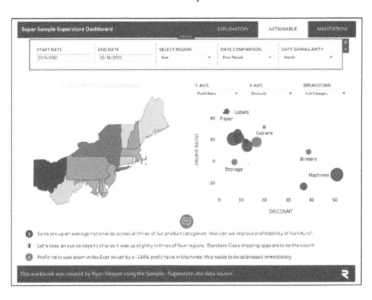

Next, in the upper-right corner of the selected container, click the drop-down menu and choose Add Show/Hide Button:

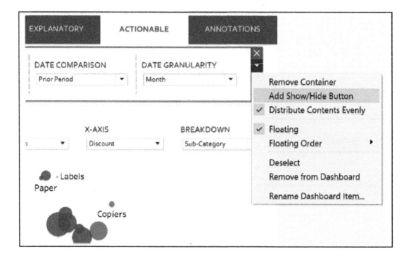

 The layout container must be floating to see the Add Show/Hide Button option.

This adds a floating object that you can place wherever you want on the dashboard. By default, the Show/Hide button displays an image of an X to indicate that the container is being shown and imply that clicking the X will close the container. When the container is hidden, the default image changes to a *hamburger* menu icon, indicating the container is closed and implying that clicking the menu will open the container.

You can test the user experience in Presentation mode by clicking the projector icon in the top ribbon. Clicking the X image will turn off the container, returning us to our cleaner design!

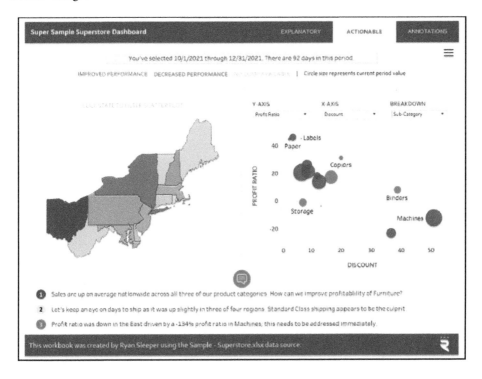

The image of the X has been changed to an image of a menu icon. Clicking the menu icon turns the container back on and the menu icon is replaced with the X button.

You can change the image settings and more by selecting the Show/Hide button object, clicking the down arrow in the upper-right corner of the object, and then selecting Edit Button.

CHAPTER 7

New Mapping Capabilities

There have been updates to map styles, layers, and even related mark types since the release of *Practical Tableau*. The first thing that you will notice is, as of version 2019.2, map styles are now a high-resolution vector format, right out of the box. Not only do the map tiles in the background look sharper, the panning experience is much more fluid compared to previous versions of Tableau Desktop.

Here is a default map style showing where all 5,004 orders in the Sample – Superstore sample data were placed:

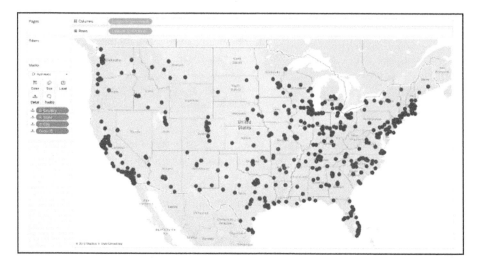

There are also three new map styles available. In addition to the traditional Light, Normal, and Dark styles, authors can now use Streets, Outdoors, and Satellite styles.

To access these layers, on the menu bar at the top of the window, click Map and then click Map Layers:

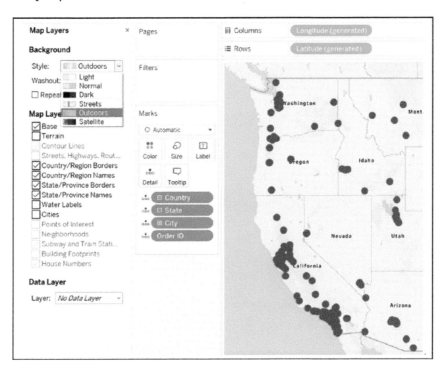

Note that similar styles were previously available via an integration with Mapbox. You can still use custom map styles from Mapbox, but the Classic tab containing 14 additional styles has been sunsetted.

Related: *Practical Tableau*, Chapter 30, "How to Make a Symbol Map with Mapbox Integration" (O'Reilly, 2018)

In addition to new map styles, there are new map layers that you can toggle on and off including Points of Interest, Subway and Train Stations, and Building Footprints. These layers are available on the same pane as the new styles and you can access them via the menu bar at the top of the window by clicking Maps > Map Layers.

I noticed that we have an unusual order pattern that runs from north to south through the state of Colorado on the first map in this chapter. Let's zoom in and turn on the Streets, Highways, Routes layer and the Points of Interest layer to see whether this reveals anything interesting:

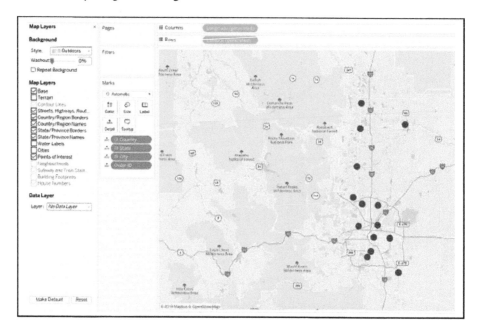

Ah-ha! There is a mountain range with several wilderness areas and national parks running through Colorado. It makes sense that all of our orders are being placed along the major interstate highway running parallel to the mountains where retail areas are more likely to be located.

To see one more mapping improvement, let's zoom back out by hovering over the map and clicking the pin icon that appears on the map controls to reset the zoom:

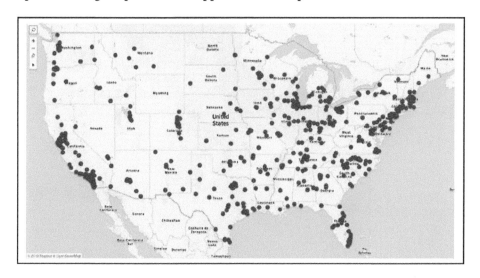

Notice that there are several areas in the country where our orders are very dense, making it a challenge to determine whether there are more orders in Southern California than South Florida, for example. When dealing with dense data points like these, there is a new mark type called Density, which you can access just by clicking the Mark Type drop-down box on the Marks Shelf.

After changing the mark type to Density, clicking the Color Marks Card will open its settings:

It turns out New York would be a better comparison point for Southern California—an insight made possible by the new Density mark type! The Density mark type is not just for maps; it is also particularly useful when you have overlapping marks on a scatter plot.

 For more ideas on how to use Tableau maps, including the new MAKEPOINT and MAKELINE functions, creating custom territories, and combining generated and custom coordinates, see the 3 Ways to Make Magnificent Maps series in Chapters 28 through 30.

CHAPTER 8

Explain Data

Tableau 2019.3 saw the introduction of Explain Data, which provides AI-driven context for your data that's available with the click of a button! To access the Explain Data insights, hover over a data point of interest and then click the light bulb icon that appears along with the tooltip command buttons. Tableau claims to analyze hundreds of explanations in the background and presents what it believes to be the most relevant when you open the insights.

To illustrate, let's finally investigate what's happening with the Sample Superstore's highest selling yet unprofitable customer, Sean Miller. As you can see on the following scatter plot of all 793 customers in the database, Sean is an outlier on the x-axis, which is great, but he's below the zero line on the y-axis for Profit:

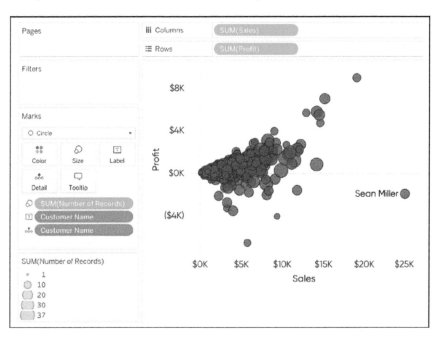

For more context on what might be causing this dimension member to perform this way, click the data point and then click the light bulb icon. Alternatively, you can right-click the data point and then, on the menu that opens, choose Explain Data:

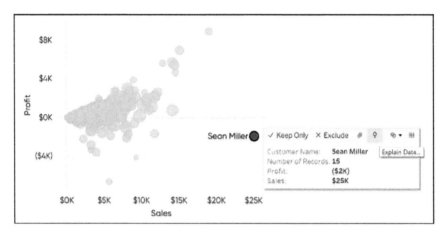

After briefly analyzing the data, Tableau displays the performance for the dimension member along with some possible explanations:

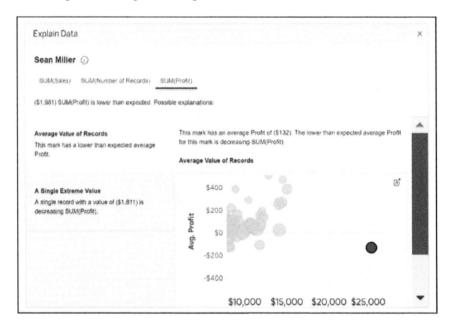

One of the most useful features of Explain Data is that it can sift through thousands of records and instantly uncover extreme values that might be leading to unexpected results. In Sean Miller's case, a single unprofitable value of ($1,811) is driving his performance down:

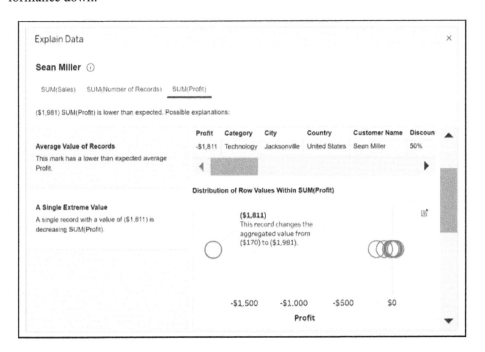

This assistance from Explain Data has helped reveal an outlier that we should examine further or a possible data entry error. It says that this single value has changed the Profit performance from ($170) to ($1,981). My first instinct is to think that somebody accidentally typed an extra *1* at the end of ($181) because that value is so close to the average for the rest of the values.

However, Explain Data also shows you the full table of fields for the data point. We can now see that this order does, in fact, appear to be an unusual case in which we provided a 50% discount on a large Technology order. It looks like Mr. Miller benefited from a great deal on a video-conferencing unit—lucky guy!

Further down in the Explain Data interface, we can see what the visualization would look like if we were to exclude the extreme value:

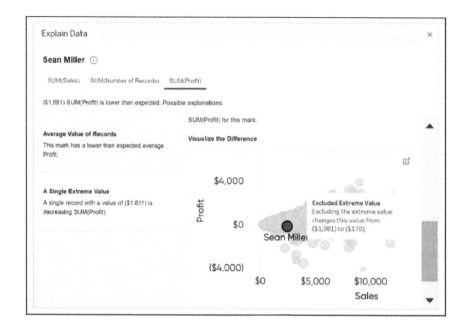

In this case, the customer is still unprofitable and worth keeping an eye on, but they're also not having nearly as big of an impact as it previously seemed. If you would prefer to move forward with this version of the chart, you can open it as a new worksheet by clicking the icon in the upper-right corner of the new chart:

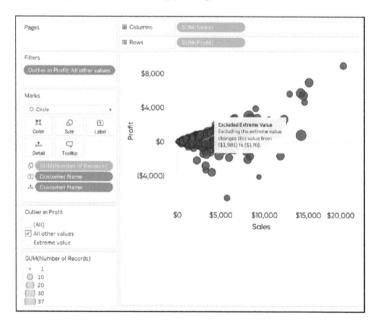

From here, you can toggle the extreme value on and off, delete the generated annotation showing you which value changed, and/or use the new version just like you would any other worksheet. Explain Data just helped us instantly understand our biggest outlier and provide a clearer picture of what's happening in our business!

Although *Innovative Tableau* is focused on the Desktop product, I would be remiss if I did not at least mention Explain Data's counterpart, Ask Data, available on Tableau Server and Online as of version 2019.1. With Ask Data, simply typing questions about your data generates visualizations to help answer those questions. It also provides the ability to include synonyms for popular phrases in your business to help with natural language processing (NLP).

With both Ask Data and Explain Data, it has never been easier to uncover starting points for deeper analyses that lead to valuable actions.

CHAPTER 9

Viz in Tooltip

One thing I like most about Explain Data is that it *visualizes* the explanations to help us reduce the time to insight. Essentially, it creates visualizations about the visualizations. Well, as of Tableau 10.5, you can display a visualization (or anything for that matter) within a visualization's tooltip! With the Viz in Tooltip feature, you can now easily insert entire worksheets into a tooltip by clicking the Tooltip marks card and the Insert button.

This feature is huge for all three of my tenets for good visualization: reducing the time to insight, increasing the accuracy of insights, and improving engagement. Just to name a few applications, it will help save valuable real estate, provide a means for revealing underlying data (see Chapter 76), or you can even use it for company branding (see Chapter 55).

How to Add a Visualization to a Tooltip

To explore this, let's look at my favorite chart, the noble bar chart, which shows us sales by region:

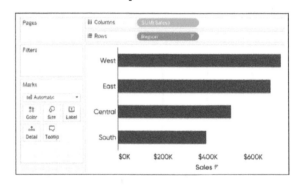

This is a great chart. We're able to use the preattentive attribute of length to instantly see insights like the West region is leading the way, and the South is lagging behind the other three regions. The chart is also sorted in descending order, making it even easier for us to compare rank order from top to bottom.

However, this chart is very explanatory, or descriptive, in nature. It's explaining what happened in the business but provides no context on how we got here or what we might want to keep an eye on.

So, to understand how sales are trending, I also created this line graph on a separate worksheet:

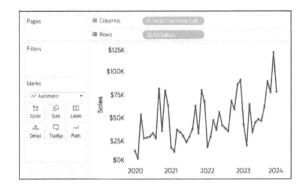

This is also a nice chart on its own, showing us that we have some seasonality in our business and we're trending in a positive direction overall.

Even though both charts are providing their own unique insights, they would be even better if they were combined so that we could view them in context of each other. Sure, we could add both charts to a dashboard and link them together by using dashboard actions, but that would take up valuable dashboard real estate that could be prioritized for other key visuals.

Related: *Practical Tableau*, Chapter 56, "Three Creative Ways to Use Dashboard Actions" (O'Reilly, 2018)

Now, with Viz in Tooltip, we can add the line graph directly to the tooltips of the marks on the bar chart. If you hover over a specific bar, the trend will appear and be filtered to the dimension member associated with the bar!

To add a visualization to a tooltip, go to the worksheet for which you want to create the tooltips and then click the Tooltip Marks Card:

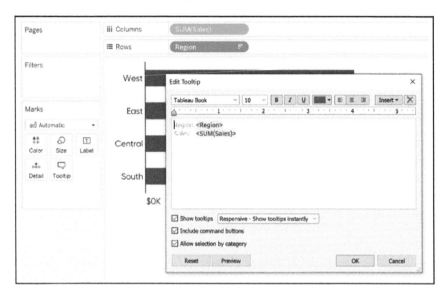

By default, the tooltip shows whatever fields are on the view. So far, we have only the Region dimension and Sales measure on the view, so those are the fields represented in the tooltip. To add the line graph visualization to this tooltip, navigate to wherever you want the visual displayed (i.e., the third line in this tooltip) and then, in the upper-right corner of the dialog box, click Insert, hover over Sheets, and choose the sheet that you want to insert:

This adds some code to the tooltip within brackets and shaded gray, indicating the values being displayed will be dynamic and based on the mark over which you're hovering. There are a few options you can overwrite in this code, including the sheet name, its maximum dimensions within the tooltip, and which fields you want the tooltip to filter on.

The default dimensions of a Viz in Tooltip are 300 pixels × 300 pixels. Because our line graph has a horizontal orientation, I will bump the max width up to 600 pixels and max height up to 400 pixels.

The default filters for a Viz in Tooltip are set to All Fields. Our bar chart has only two fields, Region and Sales, and I want the trend to update for each region that is hovered over, so let's leave this setting alone:

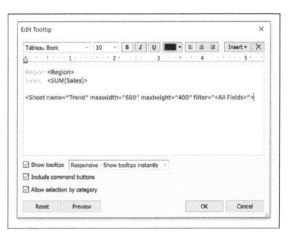

Now, hovering over the bar for any region will display the monthly trend for that specific region:

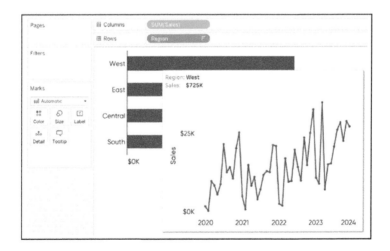

Even though the West region is leading the way in our business, it looks like sales performance has lost its momentum in recent months and is something we should investigate. This insight leading to an action is made possible with the Viz in Tooltip feature!

CHAPTER 10

Tableau Extensions

The new features covered to this point have undoubtedly taken Tableau Desktop and the value it provides to the next level since the release of *Practical Tableau*. Well, the pace of innovation will only prove to accelerate in coming years with the addition of the Extensions API. The new API allows third-party developers to code their own features that can be added to Tableau dashboards in the form of *dashboard extensions*. You can think of these as plug-ins that add additional functionality to the default dashboard features.

To see the extensions that are currently available, who developed them, and/or to download them for use, visit the Extension Gallery (*https://oreil.ly/gIs3n*):

Tableau extensions were introduced in version 2018.2, but as of version 2019.4, there are two different types of extensions with varying levels of security: sandboxed extensions and network-enabled extensions.

Sandboxed extensions are publicly hosted by Tableau and prevent any network calls outside of Tableau. Network-enabled extensions, on the other hand, can be hosted by third parties other than Tableau, meaning that your data can be sent anywhere! It is important to be aware of where your data is being sent, particularly when working with sensitive company data. To determine whether it is sandboxed, just look for the icon with the image of a cloud and lock in the upper-right corner of each extension's card.

One of my favorite things about dashboard extensions is they have become a sandbox (no pun intended) for Tableau to explore new features and provide a means for authors to become familiar with the new functionality before they are officially released in the product. For example, parameter actions and the long-awaited dynamic parameters were both available as extensions long before making it into the Desktop product.

To use a dashboard extension, download and install it from the Extension Gallery. These files end with the extensions *.trex*, and it's good practice to save these in your Tableau Repository at Documents > My Tableau Repository > Extensions so that you can keep track of them later.

When you are ready to use the extension on a dashboard, simply drag an Extension object onto the dashboard from the Dashboard pane, choose the extension, and then configure it.

Tableau is on an aggressive release schedule; new features are introduced every three months. Keep an eye on the web page for additional new features (*https://oreil.ly/NdP_K*), including visualization animations, dynamic parameters, and much more.

The next three sections of *Innovative Tableau* will show you how to use the new features shared in this section in real-world applications and present many tricks for taking existing features to the next level!

PART II

More Chart Types

CHAPTER 11

Chart Types Introduction: Going Beyond Show Me

The partners for which Playfair Data work often react to our deliverables with, "You made that in Tableau?!" This is the biggest compliment we can receive because it means we've elevated what they thought was possible in the software. When it comes to chart types, this is typically achieved with attention to detail to Tableau's formatting options, utilizing a dual axis to provide additional formatting or context, pulling in additional design elements from outside of Tableau (which we discuss in Chapter 98), and/or using innovative techniques stakeholders did not know were possible.

This section shares tricks for putting your own spin on traditional chart types, how to's for creating real-world applications, and even tutorials on creating chart types developed by me and our team, personally.

Going Beyond Show Me with the Authoring Interface Canvas and Chart Type Categories

Out of the box, Tableau comes with 24 chart types that you can immediately create just by clicking the Show Me button in the upper-right corner of the Authoring interface and making a selection. If you have the correct combination of dimensions and measures, you can literally lay the foundation for 24 chart types by clicking two buttons. But Tableau is capable of much, much more.

In fact, as soon as you realize that every chart in Tableau is simply dots plotted across two axes, there is no limit to what you can "draw" on the Authoring "canvas." If you have a measure for the x-axis value and a measure for the y-axis value, place the measures on the Columns Shelf and Rows Shelf, respectively, and the point will be plotted on the view.

Related: *Practical Tableau*, Chapter 35, "How to Make Custom Polygon Maps" (O'Reilly, 2018)

Here's how this idea was used to draw a custom polygon shape of a football player showing the cause of concussions in the National Football League:

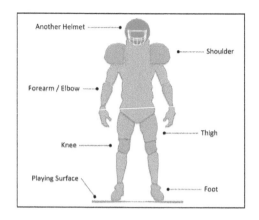

We can group chart types in Tableau into five categories:

The defaults including charts created with Show Me
> The advantage to these charts is they are extremely efficient to create, but your visualizations will look like literally millions of others.

Traditional chart types with additional formatting
> The Format pane includes formatting options for *50* different elements of each view for the entire Sheet, Rows, or Columns. This doesn't even count the ability to format the fields being used on the view for the Axis or Pane, nor does it count encoding the marks via the options on the Marks Cards. At a minimum, you should consider these options to help your traditional charts stand out.

Dual-axis(+) combination charts
> Many of the most innovative charts involve a dual axis that can be used for either formatting, analytical context, or to make the user part of the story. With the addition of the transparent sheets feature, you can now take this even further by layering sheets on top of each other.

Plotted points on x/y canvas
> I've literally seen authors recreate famous works of art in Tableau by plotting colored pixels on a view—you can think of this as a *paint by numbers* in Tableau that can draw anything. This concept, combined with data densification to fill in pixels on the view, is also what allows for custom shapes and curved lines to be used.

Mapped background images
> With the ability to integrate images in a view and float transparent sheets over images, you can bring in custom images for maps, gauges, and more.

I hope the next 35 chapters on charts across these 5 categories will help you think outside the box to take your charts up a level and inspire you to create something new.

CHAPTER 12

The Data-Ink Ratio and How to Apply It in Tableau

In his 1983 book, *The Visual Display of Quantitative Information* (Graphics Press), Edward Tufte introduced the concept of the *data-ink ratio* in data visualization. In case you are not familiar with this classic book and concept, the data-ink ratio is the portion of ink (i.e., pixels) that makes up data-information on the view.

Sometimes, to innovate, you must look back, and I've come to realize this single, decades-old concept is the core of almost all my design tips. Because the data-ink ratio is a common thread between most of the charts I create with Tableau, let's begin this section by explaining the concept, addressing its two fundamental flaws, and showing you how to apply its principles within Tableau.

Ideas for Improving the Data-Ink Ratio of Tableau Visualizations

There are several ways to improve the data-ink ratio in Tableau by changing the default format settings the software provides. To illustrate, let's improve the data-ink ratio in the following Orders by Category bar chart built with the Sample – Superstore dataset. Along the way, I point out two fundamental flaws with Tufte's vision for the concept.

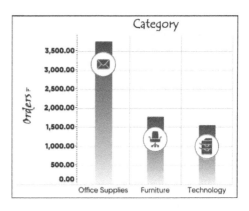

Step 1: Erase Non-Data Ink

First, we will remove non-data ink such as graphics. In the same book Tufte introduced the data-ink ratio, he also coined the term *chartjunk* for non-data graphics that clutter the view. This happens to be one of the few times my humble opinion diverges from Tufte. I believe that non-data ink such as icons can help make a data visualization more engaging so that more people stop to look at it, adopt it, and hopefully eventually act on it.

For this reason, I use a slightly different definition for chartjunk which is *any graphics that distract from the data rather than add value.*

In a 2007 study (*https://oreil.ly/E2BDX*), Inbar found that students preferred small amounts of chartjunk. Further, a 2008 study by Hockley (*https://oreil.ly/J7yMP*) found that graphics improve memorability.

The key is using graphics tastefully so that they contribute to processing and retention and don't distract from them. In my Tableau example, the icons can help the user remember which category performed best, but the varying locations of the icons within the bar imply the locations have meaning—when they don't. This increases the cognitive load on the audience and causes users to slow down to figure out what the location of the graphics on the y-axis indicates.

Here's how the view looks after removing chartjunk:

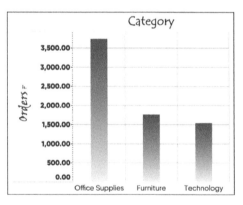

The next tips are to remove effects such as gradients:

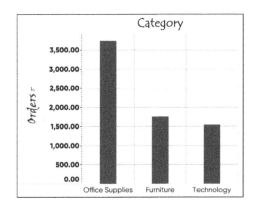

Remove some of the variation in font styles and formats:

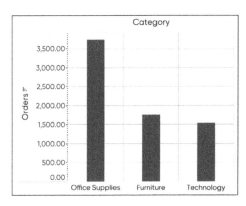

Remove the grid:

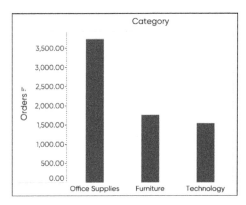

And remove borders:

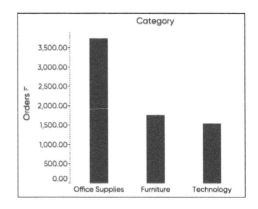

It's the last two tips that have also gar-nered criticism of the data-ink ratio. Removing grid lines and the top bor-ders makes it more difficult for users to accurately compare performance across categories. I sometimes split the difference in Tableau and add a light, dotted grid that helps the user com-pare performance but also provides a nice aesthetic:

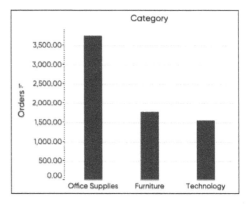

All of these formatting changes can be accessed in Tableau by right-clicking anywhere in the view and then, on the menu that opens, choose Format.

Step 2: Erase Redundant Data Ink

Just because pixels on a view convey data does not mean they are all necessary. You can also have *redundant* data ink. Some easy opportunities to remove redundant data ink include the following:

Remove extra labels and/or consolidate them into a title:

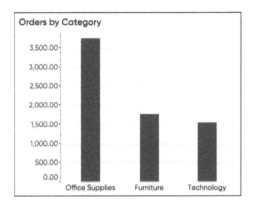

Remove unneeded decimal places and excessive axis tick marks:

To change decimal places for a specific axis, right-click the axis and then choose Format. Then, use the Numbers drop-down box to format the numbers.

To change the default number of decimal places every time a measure is used, in the Measures area of the Data pane, right-click the measure, hover over Default Properties, and then choose Number Format.

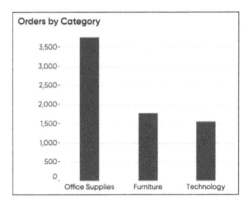

Or better yet, use direct labeling instead of axes. This works particularly well with bar charts with a relatively small number of marks on the view:

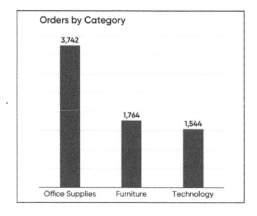

Step 3: Revise and Edit

The last step in maximizing the data-ink ratio is to revise and edit. Here is a comparison of where we started and ended up after applying the techniques listed throughout this chapter:

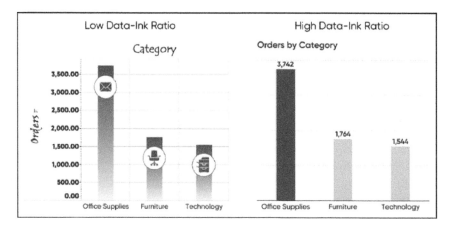

Related: *Practical Tableau*, Chapter 73, "Three Ways to Make Beautiful Bar Charts in Tableau" (O'Reilly, 2018)

CHAPTER 13

3 Ways to Make Charming Crosstabs: Tip 1

I've explained before that a spreadsheet is not a data visualization because it does not take advantage of any preattentive attributes that will help you make sense of the data. Alas, many stakeholders continue to ask for crosstabs (text tables), either so they can do their own manipulation to the data or just because they feel more comfortable seeing the raw numbers. Well, if I have to make crosstabs, I intend to make the best dang crosstabs the world has ever seen!

That's why for the first time, I'm letting the genie out of the bottle and sharing three tips for making your text tables more effective and engaging in Tableau. Over the next three chapters, we cover how to increase the number of columns in a Tableau crosstab, how to create custom table headers, and how to align table text with precision.

 Related: *Practical Tableau*, Chapter 18, "A Spreadsheet Is Not a Data Visualization" (O'Reilly, 2018)

How to Make Flexible Text Tables with Limitless Columns

Prior to Tableau Desktop 2019.4, Tableau allowed you to create a text table with six column headers before it began concatenating dimensions.

You can easily increase this column limit to 16, though: on the menu bar at the top of the window, click Analysis > Table Layout > Advanced. This opens the Table Options dialog box in which you can change the column limits from 6 to 16:

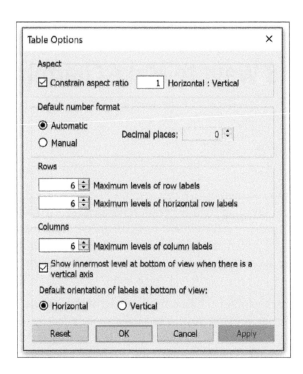

As of Tableau 2019.4, these limits have been increased to 50. If you need to make a table with 50 columns, I suspect Tableau might not be the best choice for you. In any case, if you want to go beyond 16 columns before upgrading to 2019.4 and beyond, and/or you want to build the most flexible tables possible, you can use my Placeholder hack.

To begin, make a calculated field with the following formula:

```
MIN(0)
```

This formula creates a new measure and guarantees that the value of that measure will always be 0. After you've set up this calculated field, you can place it on the Columns Shelf or Rows Shelf as many times as you want. The reason this unlocks some unique flexibility is because each measure on the Columns Shelf or Rows Shelf gets its own Marks Shelf, where you can independently edit the Marks Card. So, you can place the Placeholder field on the Rows Shelf 50 times, change the mark type to Text, and place whatever measure you want on each column on the Text Marks Card.

Check it out: here's the start of a 50-column text table broken down by the Segment and Category dimensions. I've placed the Profit measure on the Text Marks Card of the first Placeholder measure:

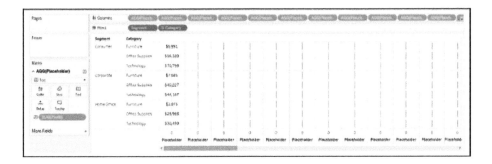

Of course, this is a very extreme example, but I wanted to show that you can already get to 50 (or more) columns, even though Tableau thought better of it and originally limited you to 16. If I ever build text tables, I typically use this approach to make tables consisting of three to five columns. What I like more about this Placeholder hack than the ability to use excessive columns is the fact that you can use a combination of mark types. For example, some cells can be text, whereas other cells can be indicator circles that change color based on performance. This is achieved by changing the mark type on each Marks Shelf.

CHAPTER 14

3 Ways to Make Charming Crosstabs: Tip 2

How to Make Custom Tableau Table Headers

One drawback with my Placeholder crosstab tip in Chapter 13 is that you lose the column headers. Fortunately, you can use the same Placeholder hack to create the headers themselves. This unlocks some interesting possibilities. For example, you can color some of the headers but not others so that you can direct the end user to focus on certain fields. Or, you could color the headers based on period-over-period performance.

To create custom headers for your text tables, start a new sheet and put the Placeholder measure on the Columns Shelf once for each column in your table. For this example, let's pretend that we're using the Sample – Superstore dataset to look at Sales, Profit, Discount, and Quantity by Category and Sub-Category with a Placeholder table. Note that I've hidden the zero lines and gridlines.

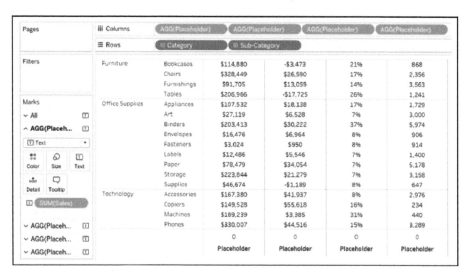

See the limitation with this table? No column headers for my measures. On a new sheet, I put the Placeholder measure on the Columns Shelf six times so that my two dimensions and four measures each get their own column and respective Marks Shelf that I can independently edit.

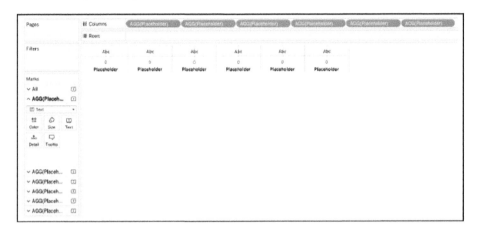

For the first two column headers, I'll leave the mark type as Text and change the column names to Category and Sub-Category, respectively. One of the ways you can use Tableau in the flow (see Chapter 61) is to directly double-click a Marks Shelf below its Marks Cards:

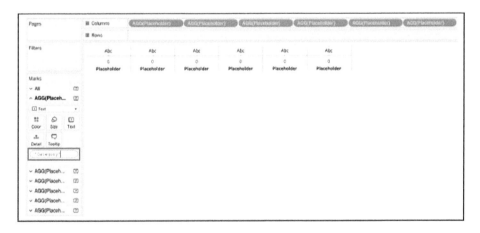

This opens a field in which you can type a string. By default, this string will be on the Detail Marks Card, but if you drag it to the Text Marks Card, the name of the column header will populate.

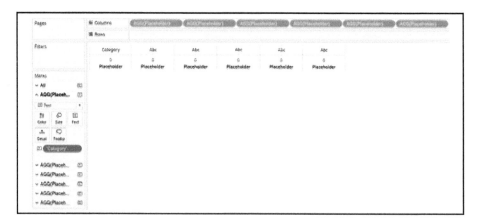

I repeat this step for the Sub-Category header. For the four remaining column headers, I change the mark type from Text to Square. For the squares to fill the entire cell, click the Size Marks Card and increase the size of the four squares. To add the names of the columns, use the approach described in the previous paragraph to add the column name in the flow, but drag it to the Label Marks Card instead of the Text Marks Card. Here's how my foundation looks at this point:

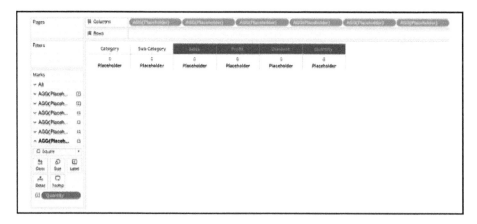

From here, there are some good options for deriving value from this custom header. One idea is that you can color the column headers (i.e., the squares) to help draw your end users' attention to one or more columns. This is accomplished by navigating to each Marks Shelf, clicking the Color Marks Card, and then choosing different colors. I've also updated the formatting to hide the Placeholder header, hide the zero lines, and add borders (column and row dividers) to the headers:

Another idea is to color the squares based on period-over-period performance for each measure. If you need help doing this, see how to make performance indicator titles in Chapter 38. Here's how my custom header looks with a month-over-month comparison:

After you have either the *focus* or *comparison* customizations built, place the header in line with the table on a dashboard.

Category	Sub-Category	Sales	Profit	Discount	Quantity
Furniture	Bookcases	$114,880	-$3,473	21%	868
	Chairs	$328,449	$26,590	17%	2,356
	Furnishings	$91,705	$13,059	14%	3,563
	Tables	$206,966	-$17,725	26%	1,241
Office Supplies	Appliances	$107,532	$18,138	17%	1,729
	Art	$27,119	$6,528	7%	3,000
	Binders	$203,413	$30,222	37%	5,974
	Envelopes	$16,476	$6,964	8%	906
	Fasteners	$3,024	$950	8%	914
	Labels	$12,486	$5,546	7%	1,400
	Paper	$78,479	$34,054	7%	5,178
	Storage	$223,844	$21,279	7%	3,158
	Supplies	$46,674	-$1,189	8%	647
Technology	Accessories	$167,380	$41,937	8%	2,976
	Copiers	$149,528	$55,618	16%	234
	Machines	$189,239	$3,385	31%	440
	Phones	$330,007	$44,516	15%	3,289

Now, not only can stakeholders see the raw data, but the colors in the headers help to clearly identify what to focus on.

CHAPTER 15

3 Ways to Make Charming Crosstabs: Tip 3

How to Align Text with Precision

Notice on the tables that are built with the Placeholder measure in Chapters 13 and 14 that the default alignment of the text is centered. What's happening is the text is aligned on a quantitative axis, but there is only one number: 0. Instead of using Text as the mark type, you can change the mark type to Gantt Bar and show the values as labels. The values will automatically move from the Text Marks Card to the Label Marks Card when you change the mark type from Text to Gantt Bar:

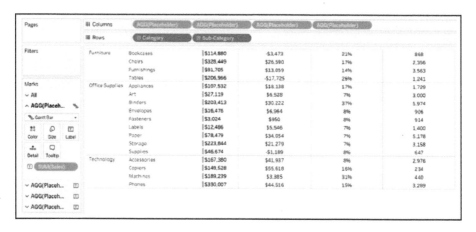

This Gantt mark always shows up at zero, which means that you can change the axis range to get the mark to align precisely where you want. I hid the Placeholder header that was showing the axis range of 0, so first I need to get that back by right-clicking any of the Placeholder measures on the Columns Shelf and then, on the context menu that opens, choose Show Header. Now that I can see the axis header, I can right-click it and choose Edit Axis.

One option in this dialog box is to fix the axis range. It's generally a good idea to right-align numbers, which I can do by setting the axis range from –5 to 0. This will move the Gantt mark to the right of the cell, making the labels appear right aligned:

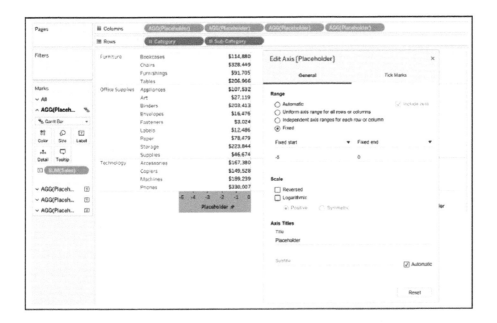

Here's my final *charming crosstab* after changing the measure mark types to Gantt Bar, right-aligning them with custom axis ranges, and hiding Placeholder headers again:

Category	Sub-Category	Sales	Profit	Discount	Quantity
Furniture	Bookcases	$114,880	-$3,473	21%	868
	Chairs	$328,449	$26,590	17%	2,356
	Furnishings	$91,705	$13,059	14%	3,563
	Tables	$206,966	-$17,725	26%	1,241
Office Supplies	Appliances	$107,532	$18,138	17%	1,729
	Art	$27,119	$6,528	7%	3,000
	Binders	$203,413	$30,222	37%	5,974
	Envelopes	$16,476	$6,964	8%	906
	Fasteners	$3,024	$950	8%	914
	Labels	$12,486	$5,546	7%	1,400
	Paper	$78,479	$34,054	7%	5,178
	Storage	$223,844	$21,279	7%	3,158
	Supplies	$46,674	-$1,189	8%	647
Technology	Accessories	$167,380	$41,937	8%	2,976
	Copiers	$149,528	$55,618	16%	234
	Machines	$189,239	$3,385	31%	440
	Phones	$330,007	$44,516	15%	3,289

We've created a text table that has limitless columns, custom headers to assist under-standing, and properly aligned numbers!

CHAPTER 16

3 Ways to Make Handsome Highlight Tables: Tip 1

To draw a highlight table in Tableau, you need one or more dimensions and exactly one measure. This is the same criteria to draw a raw text table in Tableau, except with highlight tables, you're limited to one measure instead of one or more measures. This one measure is what encodes the cells in the table by the preattentive attribute of color. It's essentially a spreadsheet with colored cells.

The highlight table is my favorite chart type for introducing the value of data visualization. I think it works well because most companies are still using spreadsheets for most of their reporting, and by converting a text table to a highlight table, the audience is forced to take advantage of at least one preattentive attribute. This kind of becomes a gateway to more complex visualizations.

Highlight tables are already more engaging and effective than a text table/crosstab view, but the next three chapters aim to provide more ways to make your highlight tables even better in Tableau.

How to Make a Highlight Table in Tableau

To lay the foundation for the next couple of tips, I'm going to make a quick highlight table in Tableau using the Sample – Superstore dataset.

There are several ways to make this, but for this example, let's preselect the fields Order Date, Sub-Category, and Profit; then, in the upper-right corner of the authoring window, in the Show Me pane, select the third option on the top row, *highlight tables*:

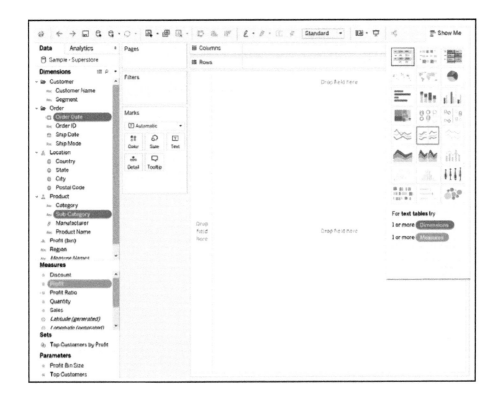

This generates the following view. As you can see, the cells are colored on a diverging spectrum with the worst profit values colored a dark orange, and the best profit values colored a dark blue:

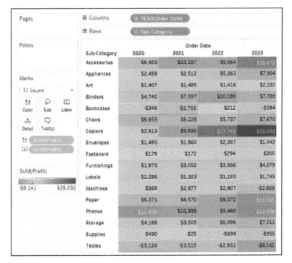

How to Format Marks for a Highlight Table

My first tip to creating more engaging highlight tables involves using the formatting options available to us in Tableau. Specifically, let's format the cells of the highlight table using the Color and Label Marks Cards. First, whenever I create a chart type and the marks are touching, I like to add a white border around the marks to provide a clean-looking separation. Examples of these chart types include stacked bar charts, stacked area graphs, filled maps, and highlight tables.

To add a white border around the marks, click the Color Marks Card, click the Border drop-down box, and then, on the color palette that opens, change the color from None to white (or a color of your choosing):

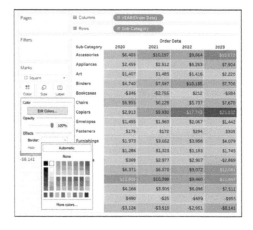

Regarding the labels, note that some of the values are colored black, whereas others are colored white. Tableau is trying to help you here by ensuring there is enough contrast between the label and the hue of the underlying color. I often opt for darker and bolder colors for highlight tables, which we discuss in the next tip, so I typically like to change all the labels to white. To do so, click the Label Marks Card, click the Font drop-down box, and then change the color to white:

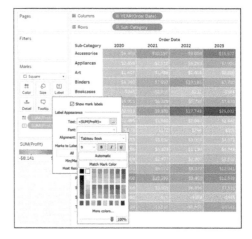

All black is also a good choice depending on the colors selected in the next tip. We've already made the highlight table cleaner with some quick changes to the default formatting options, but keep reading for more ideas on improving the design and understanding of highlight tables.

CHAPTER 17

3 Ways to Make Handsome Highlight Tables: Tip 2

Use Colors in Moderation

Depending on your objective or business question, it doesn't always make sense to use a full diverging color palette in a highlight table.

Notice in the highlight tables created in Chapter 16 that the colors for the values of −$25 and $212 are very challenging to differentiate. These numbers mean very different things to the business—one cell represents profitability and the other represents a loss—yet they are both colored gray because they are in the middle of our diverging color spectrum.

You likely will have only a few cells that are easy to differentiate depending on the range of values in your highlight table. If your business question is to truly identify the highest and lowest values in the dataset, you will benefit from reading Chapter 77. Instead, I often want to know which cells contain positive values and which cells contain negative values. This would be easier to differentiate if my highlight table were limited to two colors: one for positive values and one for negative values.

To make this change, you can either click the Colors Marks Card and then choose Edit Colors or double-click the color legend. Both paths open the Edit Colors dialog box:

Lowering the Stepped Color option to two will show one color for positive values and another color for negative values. You can also customize the colors being used on each end of the spectrum by clicking the colored square on each end of the palette and choosing a different color. Here's how my highlight table looks after reducing the steps to two and choosing two custom colors:

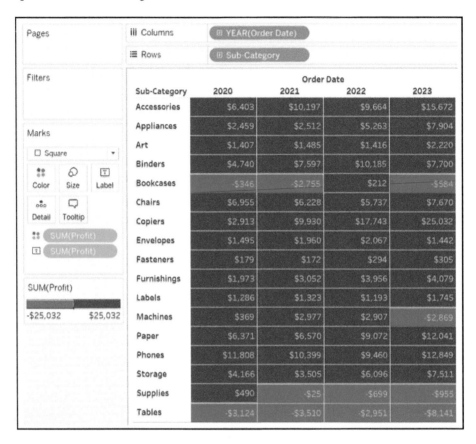

With this technique, the values −$25 and $212 mentioned earlier have been classified as red for negative and blue for positive, respectively, making the highlight table significantly easier to analyze.

CHAPTER 18

3 Ways to Make Handsome Highlight Tables: Tip 3

Color Cells Based on a Discrete Segmentation

Two colors will not be enough for every objective. The two colors used in the previous tip provide more value than a plain-text table can alone, but this approach would be most useful if you were able to color the cells based on a performance segmentation of your choosing. For example, perhaps you have a goal for each sub-category or want to compare each value to the performance during a different period. For illustration purposes, let's assume that we want to color the cells red if profit is negative, blue if profit is above our goal of $10,000, and gray for every remaining value.

To get this effect on other chart types, you would start by setting up a calculated field like this:

```
IF SUM([Profit]) >= 10000 THEN "Goal Met"
ELSEIF SUM([Profit]) < 0 THEN "Negative Profitability"
ELSE "Other"
END
```

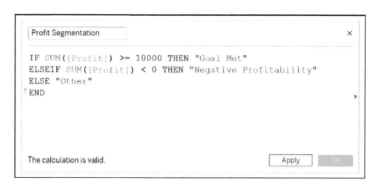

But let's see what happens when we continue building on the highlight table that we worked on in Chapters 16 and 17 and replace the Profit measure on Color Marks Card with our newly created Profit Segmentation calculated field:

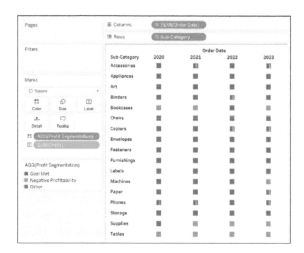

This is happening because we are now encoding the Square mark type by a discrete calculated field rather than a continuous measure. Even when we increase the size of the squares using the Size Marks Card, the highlight table doesn't look right, because the squares bleed into neighboring cells:

Fortunately, there's a trick for perfectly coloring the cells between the lines.

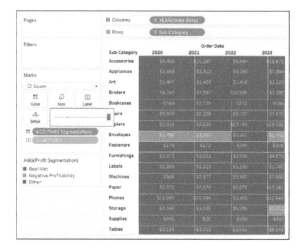

If you add a second discrete field to both the Columns Shelf and Rows Shelf, this will correct the grid. The trick is instead of choosing another dimension from the dataset or replicating the fields already on the view, we add a *blank* dimension. To do so, double-click an open area of the Columns Shelf, type two double quotation marks (""), and hit Enter:

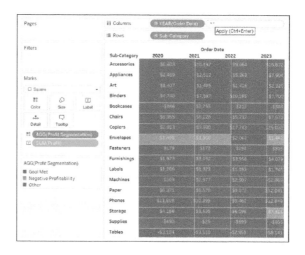

Repeat this step for the Rows Shelf. Here's how my view looks after adding a blank dimension to both the Columns Shelf and Rows Shelf and hiding their headers. To hide the headers, right-click the newly created dimensions and then, on the menu that opens, deselect Show Header:

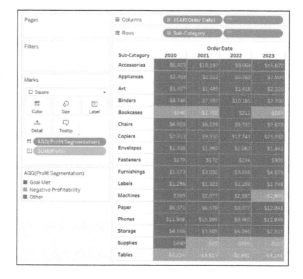

Here's how my view looks after mapping the colors to my original use case (red for negative values, blue for values above goal, and gray for everything else):

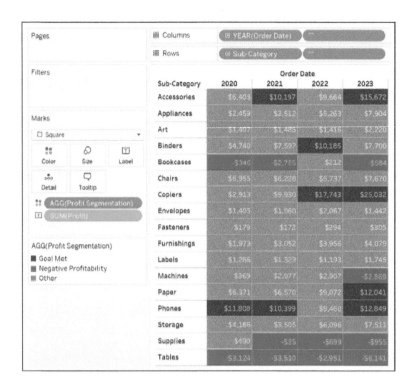

When you create a highlight table using discrete fields, the border formatting is no longer on the Color Marks Card but controlled by the Row Divider and Column Divider options on the Format pane. To access the format options, right-click anywhere on the view, choose Format, and then click the Borders tab. This gives you even more flexibility for the formatting options of the borders. You can choose not only different colors, but different line styles and weights.

Here is how my final highlight table looks after incorporating all of the tips in this chapter series and formatting the borders:

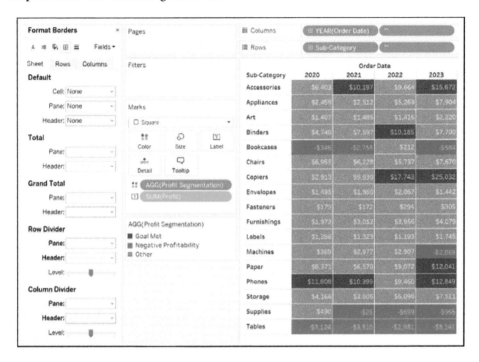

Compare this to the default highlight table from Chapter 16! We've not only enhanced a text table to introduce the value of the preattentive attribute of color, but we've made it easier to answer our business question while improving the design of the highlight table.

3 Ways to Use Dual-Axis Combination Charts: Tip 1

Dual-axis combination charts, or combo charts, are named that because they have two axes and they display a combination of different mark types. For example, you can create a visualization that displays a measure with bars on one axis and another measure as lines on the second axis. This is one of my favorite chart types to use in Tableau because the ability to add a second axis and control the axes independently of each other unlocks some additional flexibility. This newfound flexibility creates several practical applications that you can use to improve your analysis, user experience, and design.

The next three chapters show you how to make a dual-axis combo chart in Tableau as well as three different ways to use them: their traditional use; a method for making your end user part of the story; and an option for improving the aesthetics of your dashboard.

Two Ways to Make Dual-Axis Combo Charts

Let's begin by making a traditional dual-axis combination chart using the Sample – Superstore dataset. Even if this is familiar to you, I'll be sharing a second approach that you might not know and will save you a click. This first chart shows sales by year as bars on one axis and profit ratio by year as lines on the other axis. Both measures are also broken down by the Category dimension.

Let's create one of the charts, Sales by Year by Category:

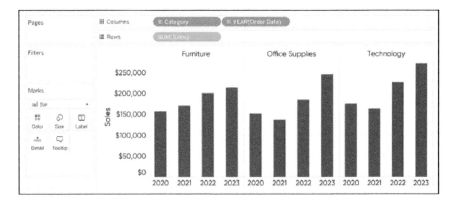

Next, place the second measure (in this case, Profit Ratio) on the Rows Shelf:

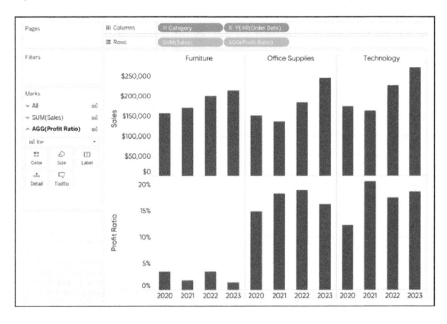

At this point, we have two individual bar charts on two rows. There are two ways in which we can convert these separate bar charts into a dual-axis bar chart. The first way, and the way most people learn, is to click the second measure pill on the Rows Shelf and then, on the drop-down menu, choose Dual Axis:

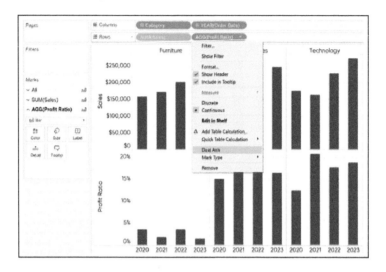

The second—and slightly more efficient method—is to hover over the axis of the second row. While hovering, a green triangle will appear in the upper-left corner of the axis. You can drag it (using the left mouse button) to the opposite axis of the left axis on the first row. When you hover over the right side of the chart, Tableau shows you a dashed line; this is where the axis will be drawn after you release the left mouse key:

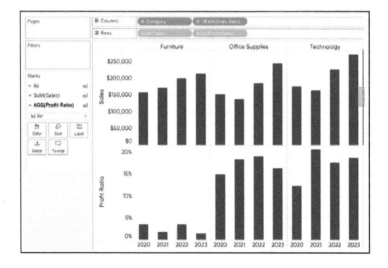

In both cases, you end up with a dual-axis bar chart:

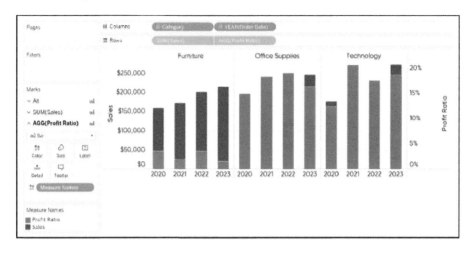

This isn't quite a dual-axis combination chart because the mark type for both charts is Bar, with the bars for the profit ratio lying on top of the bars for sales. For a dual-axis *combination* chart, we need a combination of mark types. When you have more than one measure on the Rows Shelf or Columns Shelf, each measure gets its own Marks Shelf. This means that you can independently edit the mark types for each measure.

Here's how the final view looks after I change the mark type of the Profit Ratio measure from Bar to Line:

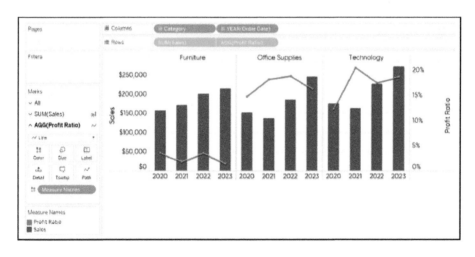

One word of caution when using the Line mark type. Lines should be used only when there is a relationship between the data points over time or if they're being connected by a continuous dimension or measure. Discrete fields are processed in their order on the Rows Shelf or Columns Shelf in Tableau. So, for my example, Sales and Profit Ratio are broken down by the Category dimension first and then the Year (Order Date) dimension second. That works, but what if we want to break the two measures down by Year (Order Date) first and then Category?

Here's how it looks:

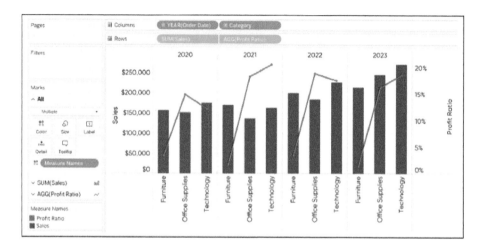

Note that we're now connecting dots with lines for data points that are not related by time. If you're ever in this situation, I suggest changing the mark type from Line to Circle to create a dot plot on the second axis:

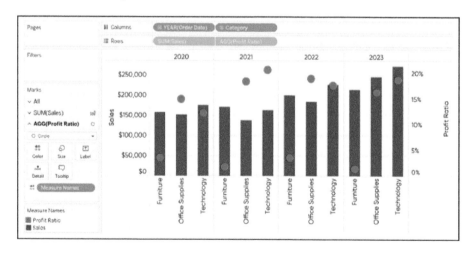

We still have a dual-axis combination chart, but we are no longer implying a relationship between the three categories.

CHAPTER 20

3 Ways to Use Dual-Axis Combination Charts: Tip 2

Use a Dual-Axis Combination Chart to Make Your User Part of the Story

For my second way to use a dual-axis combination chart, we re-create a portion of my *How does your household income rank?* Tableau dashboard (*https://oreil.ly/uyOBk*):

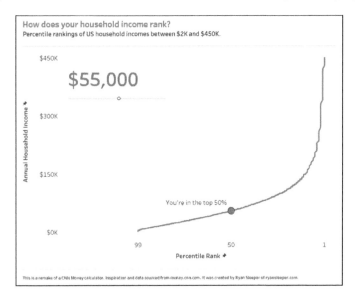

The visualization features a curve that shows how each household income ranks by percentile. This is a common *descriptive* view in that it describes high-level statistics. The real value comes by using the second axis to display where the person using the dashboard ranks on that curve. This makes the end user part of the story and provides a much more engaging user experience.

To achieve this effect, we use a parameter that would allow the end user to choose any of the household income options between $2,000 and $450,000. When the user makes a selection, in addition to the descriptive curve, the circle moves to the appropriate place based on that selection. The label also updates with a caption informing the end user of their household income percentile rank.

This is a dual-axis combination chart. The curve for the left axis uses a mark type of Line, and the circle is a second measure that only displays a circle for the end-user's selection on the right axis. The trick to getting just one circle to show up is a simple formula that computes whether the user's parameter selection matches the household income value on the Y-axis.

Here's what it looks like:

```
IF [Annual Household Income] = [User Selection] THEN [Annual Household Income] END
```

 Related: *Practical Tableau*, Chapter 14, "An Introduction to Parameters" (O'Reilly, 2018)

After we have this calculation, we can build the curve on the left axis and put the measure for the circle on the right axis:

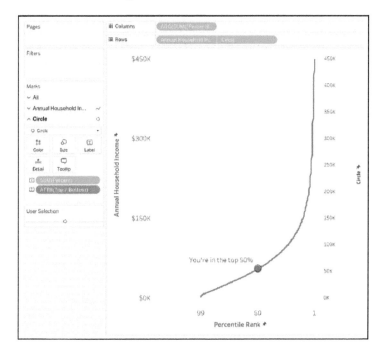

To finalize the view, we synchronize the axes by right-clicking the axis for Circle and then, on the context menu that opens, choose Synchronize Axis. This ensures that the circle will always line up perfectly with the line on the opposite axis. To finish, hide the axis on the right by right-clicking it and deselecting Show Header.

This technique of using the second axis to make the user part of the story is much more engaging than displaying the curve alone because the user can interact with the visualization by changing the parameter control, and they are able to see where they personally fall on the curve.

CHAPTER 21

3 Ways to Use Dual-Axis Combination Charts: Tip 3

Improving the Design of a Line Graph

In "Three Ways to Make Lovely Line Graphs in Tableau," Chapter 74 of *Practical Tableau* (O'Reilly, 2018), I provide a few recommendations for making traditional line graphs more engaging in Tableau. One of those approaches involves a dual-axis combination chart. This chapter shows you how it's done.

Suppose you have a line graph showing a sales-by-month trend:

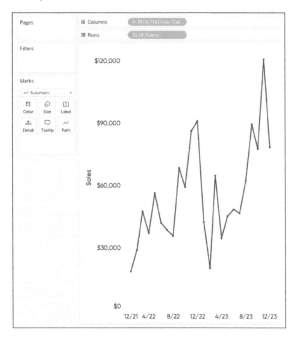

Now, let's put the Sales measure on the right axis, synchronize the axes, and change the mark type for the second axis to Area:

At this point, we have a dual-axis combination chart with sales by month as a line graph, and sales by month as an area chart.

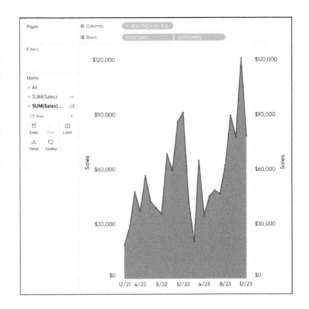

To finalize the view, we hide the right axis and reduce the opacity of the area to 10%:

You can use the second axis for only one purpose at a time, but if you are not using it for one of the applications outlined in the previous two chapters (or anything else), this third application can be an easy way to enhance your traditional line graphs. You can use this same dual-axis technique to format line markers by using a mark type of Circle on the second axis. Because the second axis has its own set of Marks Cards, you can independently format them from the traditional line graph on the opposite side!

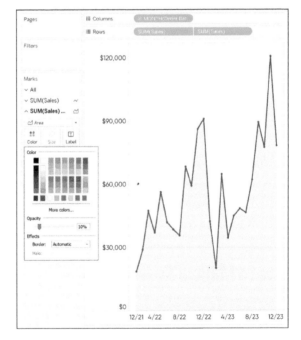

CHAPTER 22

3 Ways to Make Stunning Scatter Plots: Tip 1

When it comes to my favorite chart types, scatter plots are a close third behind bar charts and line graphs. In several industries, and especially scientific journals, scatter plots are *the* favorite choice because of their ability to reveal and communicate correlations. Another benefit of this chart type is that it is one of the few visualizations that allow you to view many marks in a small space. No, you cannot analyze every individual mark, because they will likely overlap, but scatter plots make it easy to identify outliers and the aforementioned correlations.

But wait...there's more! Because of the way scatter plots are set up with a measure on each axis, adding reference lines for the average of each axis creates a natural four-quadrant segmentation. This is a great technique for isolating different groups so that you can act on them individually.

The next three chapters show you how to make scatter plots and take them to the next level in three ways:

- How to use a formatting trick to make your scatter plots stand out
- How a calculated field can automatically break your dimension members into four usable segments
- How to make connected scatter plots

How to Make a Scatter Plot in Tableau

Scatter plots are created with two to four measures and zero or more dimensions. The first two measures form the y-axis and x-axis; then you can use the third and/or fourth measures, as well as dimensions, to add context to the marks.

For all three of the following tips, we start with this default scatter plot comparing profit ratio and sales by sub-category in the Sample – Superstore dataset:

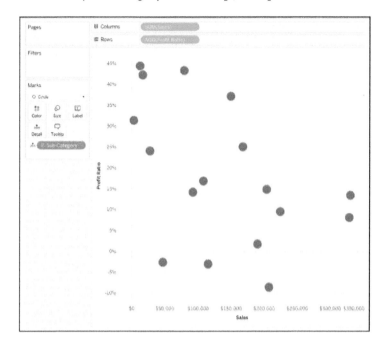

I built this chart by putting my dependent measure of Profit Ratio on the Rows Shelf to form the y-axis, and my explanatory measure of Sales on the Columns Shelf to form the x-axis. I then put the Sub-Category dimension on the Detail Marks Card to make my analysis more granular and changed the mark type from Shape to Circle.

Scatter plot is the default chart type when starting with two measures, so I could have started this chart by simply double-clicking the Profit Ratio measure from the Measures area of the Data pane and then double-clicking the Sales measure, also from the Measures area of the Data pane. Because it is best practice to put the dependent measure on the y-axis, and the explanatory measure on the x-axis, Tableau places the first measure you double-click on the Rows Shelf, and the second measure you double-click on the Columns Shelf.

You can immediately see a correlation showing that as profit ratio increases, sales tend to decrease, but we're going to keep building on this scatter plot to make it more engaging and functional.

A Formatting Trick for Better Mark Borders

As I mentioned, one of the benefits of scatter plots is that you can evaluate a lot of marks in a small space. However, this often creates overlapping marks. This is easy to alleviate by clicking the Color Marks Card and reducing the opacity of the marks, but it doesn't work well with mark borders. Borders are an effect on the Color Marks Card that I often like to add, but they do not inherit the opacity of the underlying mark, and you can also pick only one color for all marks.

The following trick allows you to control the marks and their borders independently, unlocking a great deal of formatting flexibility. To help demonstrate, I've added the Category dimension to the Color Marks Card of our default scatter plot and made the marks slightly larger:

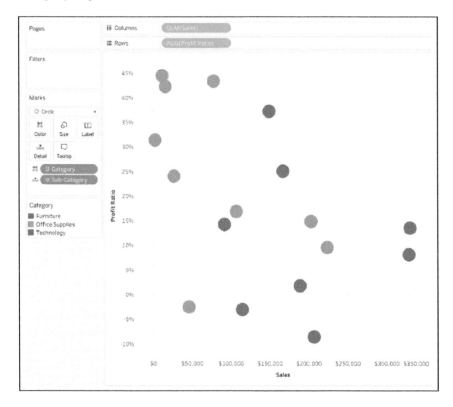

To format marks and their borders independently, place the dependent measure on the Rows Shelf next to itself a second time. This creates the same chart on two rows. Convert the two rows to a dual-axis scatter plot by clicking the second occurrence of the measure on the Rows Shelf and then clicking Dual Axis:

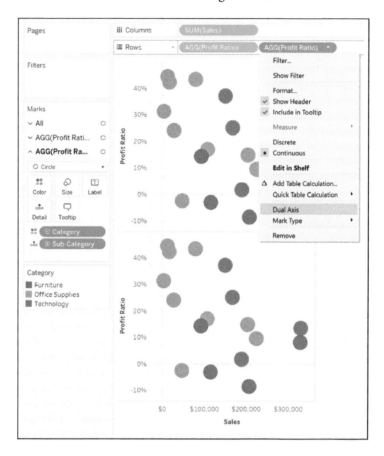

Lastly, ensure that the dual axes are synchronized by right-clicking either y-axis and then choosing Synchronize Axis. Both scatter plots are now lying directly atop the other. What's important here is now that we've added a second measure to one of the shelves, they each have their own set of Marks Cards, which you can independently edit. This means we can do things like reduce the opacity of the original marks and change the mark type of the second set of marks to Circle. This is the same technique we used to format a line or area graph in Chapter 21.

Check it out: borders that match the color of the underlying mark!

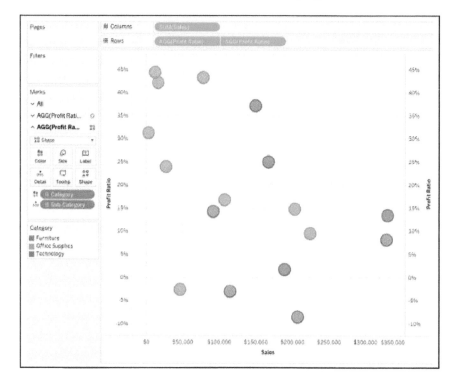

We obviously no longer need to see the second y-axis, so you can hide that by right-clicking it and then deselecting Show Header.

This first tip helps add some unique design polish to make your scatter plots more engaging. Keep reading to learn how to make scatter plots more functional and how they can even do some of your analysis for you!

CHAPTER 23

3 Ways to Make Stunning Scatter Plots: Tip 2

How to Automate Scatter Plot Segmentation

I mentioned in Chapter 22 that adding reference lines for the average of each axis creates a natural four-quadrant segmentation. Next, I share with you a way to make that segmentation permanent through a calculated field. This allows you to isolate the four segments so that you can evaluate and act on them individually based on their behavior.

To begin, on the Analytics pane, add an average reference line to each axis by dragging Average Line onto the view:

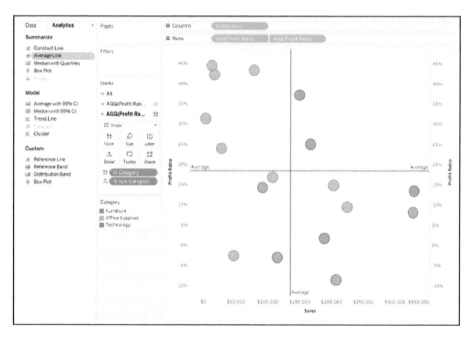

Depending on the number of axes on the view, dragging Average Line from the Analytics pane to the view can be a little confusing. This is because you must choose both the axis and the scope of the reference line. If you prefer, you can also add reference lines by simply right-clicking each axis and then, on the menu that opens, choosing Add Reference Line.

Notice we now have four quadrants on the view. The upper-left corner are the subcategories with a high profit ratio and low sales; the upper-right are those with a high profit ratio and high sales; the lower-right are those with a low profit ratio and high sales; and the lower-left are those with both a low profit ratio and low sales. Depending on the dimension members in your analysis, you will likely want to act on these four groups in different ways.

To isolate each group, you could select the dimension members in each quadrant by dragging a box around them and creating sets. A better method would be to write a calculated field to match the four-quadrant segmentation being shown and then place the calculated field on the Color Marks Card. This has several advantages:

- It makes it apparent to you and your end users what segmentation you're evaluating.

- It guarantees that you are placing the correct dimension members into the correct sets (which can be difficult if some dimension members are close to the reference lines).

- It provides a means for filtering for easier set creation.

 Related: *Practical Tableau*, Chapter 64, "Allow Users to Choose Measures and Dimensions" (O'Reilly, 2018)

As always, there is more than one way to accomplish the same objective in Tableau, but I think the easiest approach in this use case is to create a calculated field using the WINDOW_AVG table calculation. Here's the formula for our four-quadrant segmentation:

```
IF [Profit Ratio] > WINDOW_AVG([Profit Ratio]) AND SUM([Sales]) <
WINDOW_AVG(SUM([Sales])) THEN "High Profit Ratio & Low Sales"
ELSEIF [Profit Ratio] > WINDOW_AVG([Profit Ratio]) AND SUM([Sales]) >
WINDOW_AVG(SUM([Sales])) THEN "High Profit Ratio & High Sales"
ELSEIF [Profit Ratio] < WINDOW_AVG([Profit Ratio]) AND SUM([Sales]) >
WINDOW_AVG(SUM([Sales])) THEN "Low Profit Ratio & High Sales"
ELSE "Low Profit Ratio & Low Sales"
END
```

You can also use this calculation to parameterize the measures being used and/or give the segments aliases that are relevant to your business (i.e., rising stars, cash cows, dogs).

After we have the calculation, we place it on the Color Marks Card:

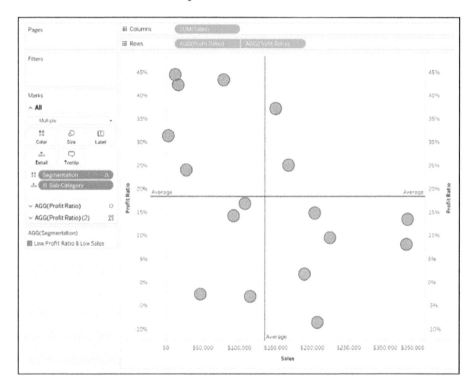

As you can see, at this point, all the marks have the same color. This is because, by default, table calculations (including WINDOW_AVG) are computed from left to right. We need to change the addressing of the table calculation from the default, Table (across), to the level of detail we are segmenting: Sub-Category.

To do this, click the measure with the delta symbol (Δ). On the drop-down menu that opens, hover over Compute Using and then choose Sub-Category:

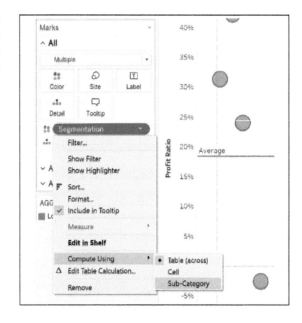

Here's how the view looks after changing the addressing and choosing colors for each of the four segments:

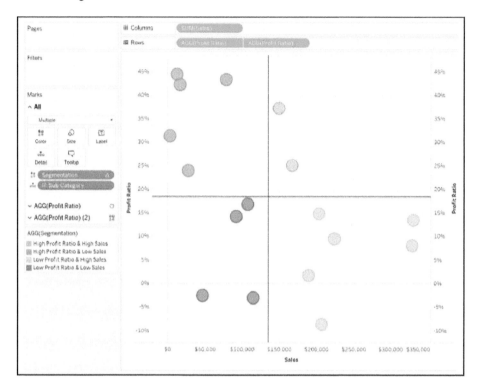

Related: *Practical Tableau*, Chapter 13, "An Introduction to Table Calculations" (O'Reilly, 2018)

In a real-world business context, we might want to handle these product segments differently. Remember, you can use this segmentation with any dimension relevant to your business, so you might be segmenting different regions, marketing channels, or customers. From here, it is very easy to isolate a segment using just two additional steps.

Suppose that we see an opportunity to improve our bottom line by focusing on the High Profit Ratio & Low Sales segment. These sub-categories are not selling much, but they're very profitable when we do sell them. Perhaps we can do a promotional campaign to drive sales to the highly profitable products, or maybe we just want to further evaluate what's making them profitable versus the other low-selling segment.

First, filter the view. On the color legend, right-click the segment of interest and then choose Keep Only:

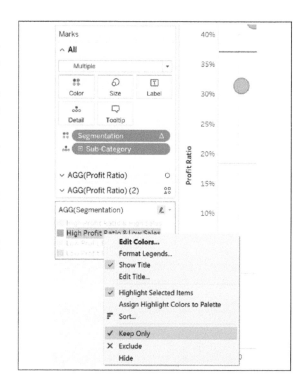

This leaves only the segment of interest on the view. Next, drag a box around the marks, right-click any one of them and then choose Create Set:

After giving the set a name and clicking OK, you will have an isolated set of the dimension members in your segment over in the Sets area of the Data pane.

For some ideas on using this, see "An Introduction to Sets," Chapter 15 of *Practical Tableau*.

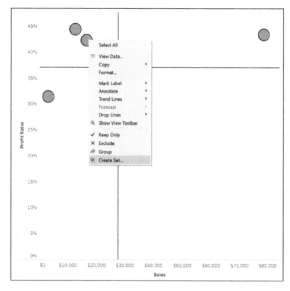

CHAPTER 24

3 Ways to Make Stunning Scatter Plots: Tip 3

How to Make Connected Scatter Plots

As I mentioned in Chapter 22, the scatter plot is already my third favorite for its innate ability to visualize many records at once, reveal correlations and outliers, and create a natural four-quadrant segmentation. One weakness of scatter plots, however, is that they don't clearly show how a dimension member's position is changing over time. Fortunately, Tableau's flexibility allows us to go far beyond the defaults and Show Me options and, in this case, can help us to literally connect the dots on a scatter plot.

With connected scatter plots, it is much easier to compare the paths of dimension members as their respective intersections on the x- and y-axes change from period to period. This chapter shows you how to make connected scatter plots in Tableau and how to use a dual axis to make the visualization more engaging and display the point order of each mark.

By the end of this chapter, you will be able to make a scatter plot that connects marks by year and is colored by any dimension of your choosing. You will also be able to communicate the order of the marks by using a table calculation on the Label Marks Card:

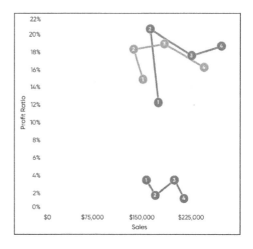

You can use the following technique with any fields, but for this tutorial, we lay the foundation of a scatter plot by using the Sample – Superstore dataset with Sales on the Columns Shelf, Profit Ratio on the Rows Shelf, and Category on the Color Marks Card:

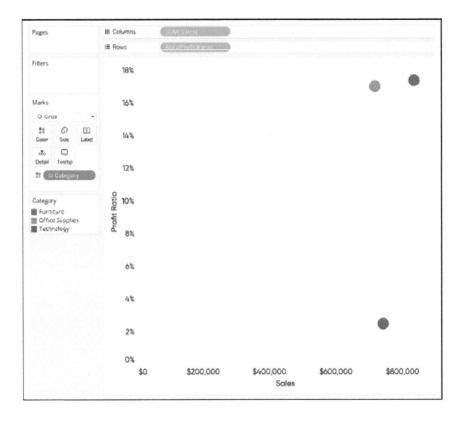

To connect the dots, change the mark type to Line:

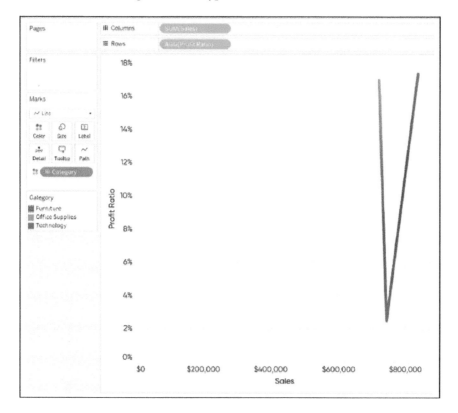

Related: *Practical Tableau*, Chapter 7, "Discrete Versus Continuous" (O'Reilly, 2018)

Next comes the most important step for creating connected scatter plots in Tableau. By default, Tableau is connecting these dots between the dimension members in the dimension on the Color Marks Card, but we can change how it connects the dots by placing a dimension on the Path Marks Card.

If we want to connect the dots by Year([Order Date]) to visualize how each Category changed position from year to year, place the Year([Order Date]) dimension on the Path Marks Card. By default, the Year date part is displayed in chronological order, but just in case, note that I used the Year([Order Date]) field as continuous:

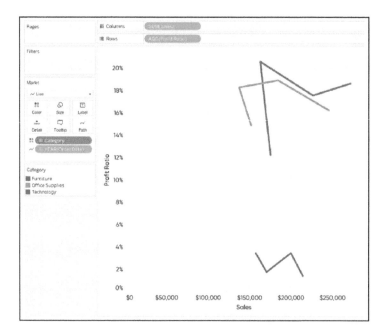

At this point, we have a connected scatter plot, and you have the option to stop here. I would at least suggest turning on the markers, which is an effect you can find by clicking the Color Marks Card:

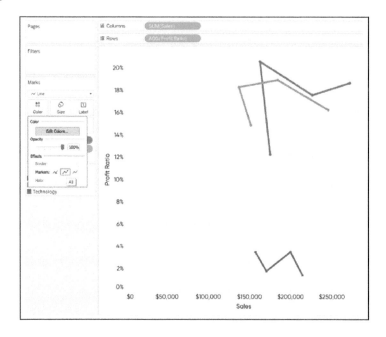

But let's keep going and make this even better by providing a visual cue that communicates to our users the order in which they should read the marks. After all, this book is called *Innovative Tableau*, not *Minimum Required Tableau*. First, let's duplicate the Profit Ratio pill on the Rows Shelf:

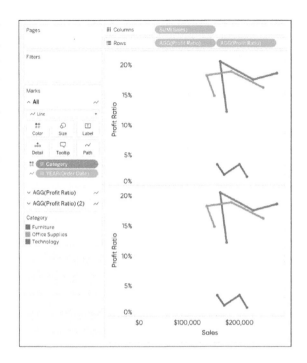

Now that there are two measures on the Rows Shelf, they each get their own Marks Shelf, which we can independently edit. Let's change the mark type for the second row to Circle so that we get a nice big mark that we can label:

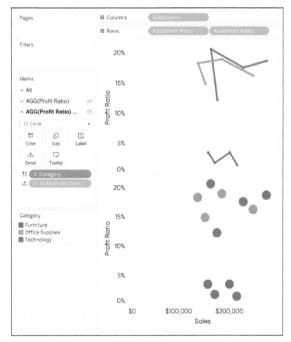

Next, convert these separate rows into a dual-axis combination chart by clicking the second pill on the Rows Shelf and then, on the menu that opens, choose Dual Axis. The connected and unconnected version of the scatter plot should be sharing the exact same scale, so as soon as they're on a dual axis, ensure that the scales are lined up by right-clicking either axis and then choosing Synchronize Axis. This yields the following dual-axis combination chart:

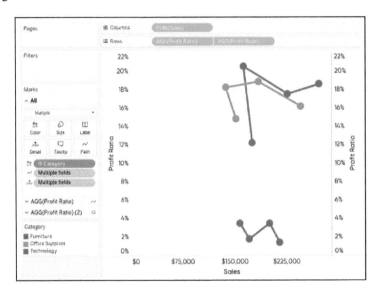

We can still independently modify the marks, so we can add a Mark Order label to the Label Marks Card of the circles only. For Tableau to display the proper mark order, start by creating a calculated field that contains only the INDEX( ) function:

INDEX( ) is a table calculation, and I think of it as synonymous with row number. If we can get the addressing and partitioning correct, we can display the proper order in which our users should read the connected marks in our scatter plot.

The first thing we see after placing this newly created Mark Order calculated field on the Label Marks Card is the number 1 on each mark because we haven't yet changed the addressing:

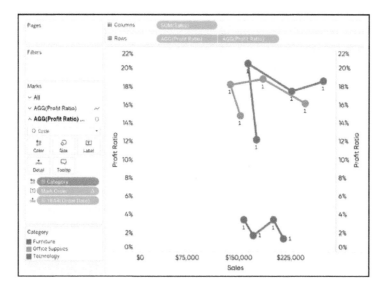

To change the addressing, click the pill with a delta symbol (indicating that there's a table calculation taking place) and then, on the menu that opens, hover over Compute Using. We want to display the proper index, or in this case, mark order, for the Order Date dimension:

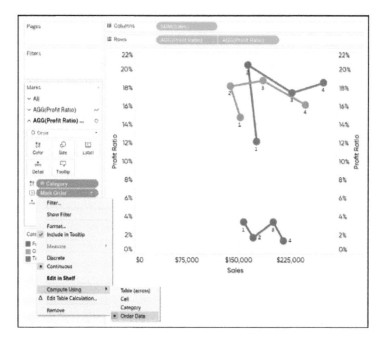

From here, all that remains is formatting! Here's how the connected scatter plot looks after hiding the right axis, centering the labels, and polishing up lines and borders:

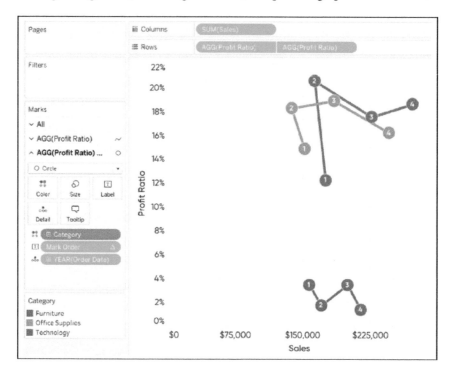

We now have a connected scatter plot with lines that imply which marks are related and also explicitly communicates how those marks have moved over time!

CHAPTER 25

3 Ways to Make Splendid Slope Graphs: Tip 1

I always say that one specific chart type per analytics scenario will help you capture roughly half the insight. Comparing categorical data? I always start with a bar chart. Want to analyze many data points at the same time and look for correlations? A scatter plot is your best bet. Trending something over time? A line graph will certainly help. *This is not to say that you should use only these three charts.* Alternatives can help keep your audience engaged, add context to your analyses, and reveal new insights.

Slope graphs, which keep only the first and last marks on a line graph, are one of my favorite alternatives for visualizing time. The next three chapters show you how to make slope graphs in Tableau and provide three tricks to improve their formatting and user experience.

How to Make Slope Graphs in Tableau

First, let me explain what a slope graph is and show you a manual way to create them. A slope graph is a line graph that connects dimension members across just two points. To illustrate, consider the following line graph looking at Profit Ratio by Region in the Sample – Superstore dataset (filtered to 2023):

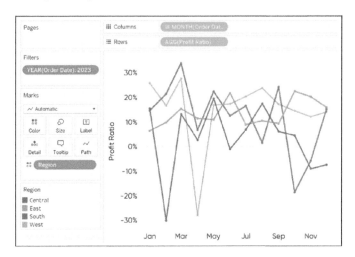

Even though line graphs are my second favorite chart (after bar chart) because of their strengths in analyzing trends over time, there are certain insights that are challenging to determine with this chart alone. For example, we have some "spaghettiing" going on from about May to September, making it difficult to analyze the trends of individual dimension members during that period. It's also tricky to analyze the overall growth or decline of each dimension member at a glance, especially with the South region, which appears to finish the year in the same place it began.

For the latter insight, slope graphs would reduce the time to insight and increase the accuracy of insight by focusing the analysis on just the first and last data points. To create a slope graph manually, simply draw a box around the points between the start and end of the date range, hover over the marks, and then, on the panel that opens, click Exclude:

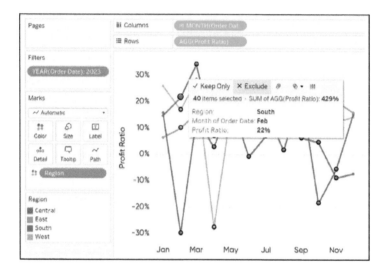

After filtering out the marks between the start and end of the range, what remains is a slope graph showing how the profit ratio has changed from January to December:

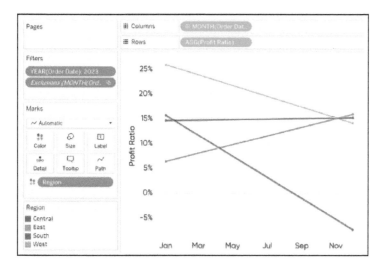

I prefer to create line graphs with continuous date fields, but with slope graphs, this leaves too many labels on the continuous axis. There are two formatting options that you can use to reduce the amount of redundant data ink (explained in Chapter 12) that the extra months are creating.

You could fix the tick marks, which we talk about in the third tip of this series, or you can change the Month of Order Date dimension from continuous to discrete. Here's how the slope graph looks if I change the Month (Order Date) field to discrete:

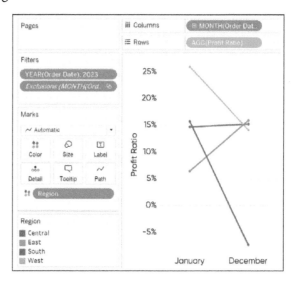

How to Make Slope Graphs Dynamic Based on a Date Range Filter

Slope graphs work great for showing how a dimension member has changed between two points in time, but they would be even better if they were tied to a date range filter. In other words, I would like for the first point and last point to automatically update based on the selection in a date range filter. I picked this trick up from Tableau Technical Evangelist, Andy Cotgreave. To illustrate, we need to take a step back and clear the filters we applied earlier in this chapter:

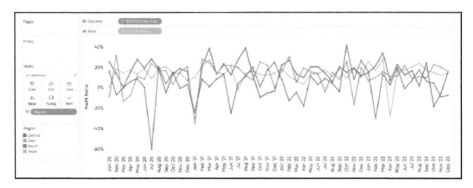

As you can see, now this is really a spaghetti graph because I removed the date filter and am now showing all 48 months in the dataset. To convert this to a slope graph this time, we set up this simple Boolean calculated field that keeps just the first and last points on the line graph:

```
FIRST() = 0 OR LAST() = 0
```

Because `FIRST()` and `LAST()` compute the difference between the current row and first or last row, respectively, a difference of zero means that the row is either first or last. `FIRST()` and `LAST()` are table calculations, so by default, they are computed from left to right.

After adding this new calculated field to the Filters Shelf, I am back to a slope graph, this time comparing the first month in the entire dataset to the last month in the entire dataset:

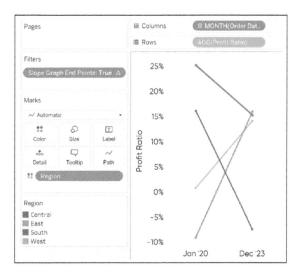

What's best about this approach is that it keeps the first month and last month, even when there is a custom date range selected in a filter. Here's how the view looks after adding a continuous date filter and choosing 12/1/2020 to 6/30/2023:

The user now has direct access to comparing the growth or decline during any date range!

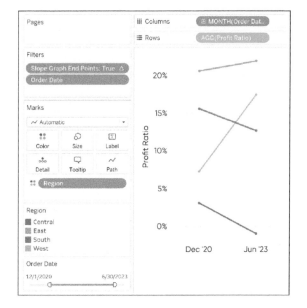

CHAPTER 26

3 Ways to Make Splendid Slope Graphs: Tip 2

How to Add a Slope Graph/Line Graph Toggle

There are times when a line graph is better for visualizing an entire trend, and others when a slope graph would be preferred to just see the overall change. Why not provide access to both options?

To toggle between a line graph and a slope graph, you begin by setting up a parameter with a data type of String. To set up the parameter, I like to right-click in any blank space in the Date pane and then, on the menu that opens, choose Create Parameter. Define allowable values to be Line Graph and Slope Graph:

Next, set up a calculated field that gives Tableau instructions for what to do when each of the allowable values is selected.

There are a few ways to write this formula, but here's one approach:

```
CASE [Line Graph / Slope Graph]
WHEN "Line Graph"
THEN NOT ISNULL([Number of Records])
WHEN "Slope Graph"
THEN FIRST() = 0 OR LAST() = 0
END
```

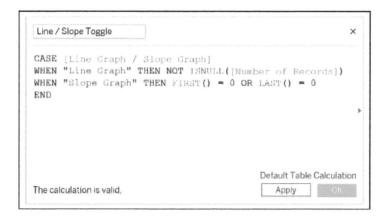

CASE [Line Graph / Slope Graph]
WHEN "Line Graph" THEN NOT ISNULL.([Number of Records])
WHEN "Slope Graph" THEN FIRST() = 0 OR LAST() = 0
END

The calculation is valid.

This is another Boolean formula, so we need the result to be True for the values we want to remain on the view. When Line Graph is selected, we're saying keep the value on the view if the record is not null (i.e., everything; so, all of the marks on the line graph). When Slope Graph is selected, we are keeping only the first and last data points as described in the previous chapter.

Now, replace the Slope Graph End Points filter from the end of Chapter 25 with the newly created calculated field. We also need to show the parameter control by right-clicking the Line Graph / Slope Graph parameter and then choosing Show Parameter Control. Now, when we select Line Graph, we see Profit Ratio by Region by Month as a line graph:

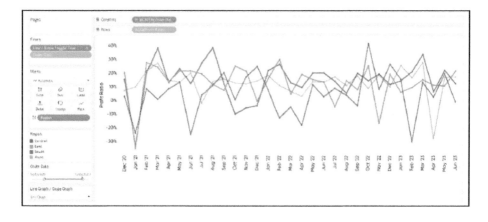

When we select Slope Graph, we see Profit Ratio by Region as a period over period slope graph:

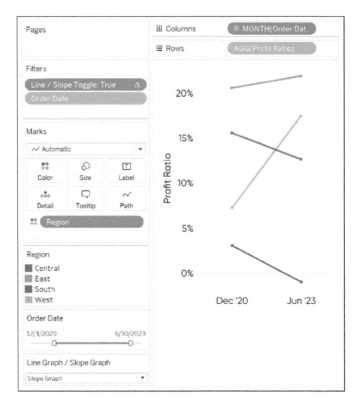

It's now up to the user whether they want to get more granular and view the entire trend over time or look at the data at a macro level to view the overall change!

CHAPTER 27

3 Ways to Make Splendid Slope Graphs: Tip 3

How to Add Vertical Lines to Connect Slope Graphs

In the last tip of this series, I share a couple of formatting hacks that allow you to draw vertical lines on the x-axis of a slope graph view. The approach is different depending on whether your x-axis is being drawn with a discrete or continuous field, so for the first example, we continue building on the slope graphs we used in Chapters 25 and 26. Note that the Month of Order Date dimension in the last image is blue, indicating that it is drawing discrete headers as columns for each month on the view.

To create vertical lines, double-click the Rows Shelf and type **MIN(1)**:

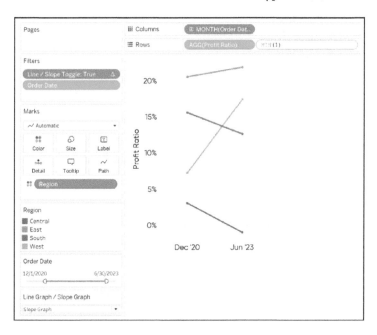

Next, remove all fields from the new row's Marks Cards, change the mark type of the second row to Bar, and convert the two rows into a dual-axis combination chart:

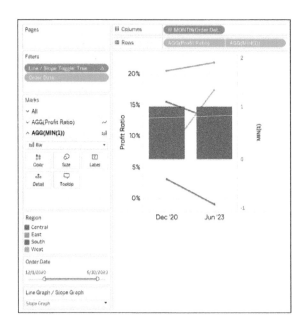

Now fix the axis range on the right by right-clicking it and then, on the menu that opens, choose Edit Axis. That opens this dialog box, where you then update the range so that it goes only from 0 to 1:

The bars are now your vertical lines! Format them to your liking by navigating to the Marks Shelf for MIN(1) and clicking the Size Marks Card and/or changing their color on the Color Marks Card.

If you would prefer the connecting lines be behind the slope graphs on the view, right-click the right axis and choose "Move marks to back." You should also hide the right axis by right-clicking it and deselecting Show Header. Here's my view after formatting.

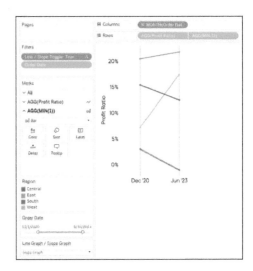

Note that these lines will show up on every month. This is ideal when formatting slope graphs with only two months as pictured, but if you are using the Line Graph / Slope Graph toggle that we set up in Chapter 26, you will see excessive vertical lines when Line Graph is selected. For best results, you can also parameterize the vertical lines so that they don't show up when Line Graph is selected in the toggle.

To do so, instead of hardcoding MIN(1) on the Rows Shelf, create a calculated field that results in MIN(1) when Slope Graph is selected in the toggle, but MIN(0) when Line Graph is selected. The code would be something like this:

```
CASE [Line Graph / Slope Graph]
WHEN "Line Graph" THEN MIN(0)
WHEN "Slope Graph" THEN MIN(1)
END
```

This field would then be used to create the bars on the second axis.

But what if I'm using a continuous field to create my slope graphs?

I mentioned earlier that discrete dates work slightly better than continuous dates with slope graphs. The reason is that discrete dates will draw just one header for the first date and one header for the second date. With continuous axes, you will see the first date, second date, and everything in between until you change the defaults.

On the plus side, continuous axes allow for more flexible formatting because you can add reference lines. To illustrate, here's how the slope graph from this chapter looks after changing the date on the Columns Shelf from discrete to continuous:

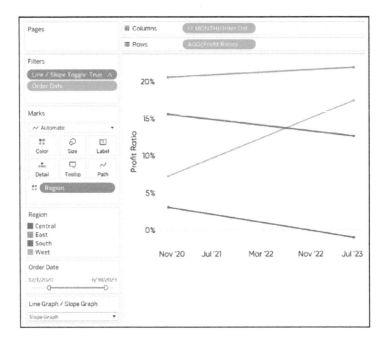

To alleviate some of the excessive tick marks and provide better padding on the sides of the chart, let's edit the axis and fix the axis range.

This approach works well for creating slope graphs when the date range is consistent. However, note that if you are using this formatting trick with a date range filter that allows the user to select date ranges with varying months, you would need to edit the axis each time to get the tick marks to line up with the line ends of the slope graphs.

Another option is to hide the x-axis all together, but you would probably want to provide some visual indication of what date range has been selected.

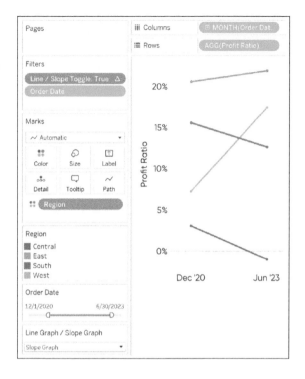

From here, to add the vertical lines that connect the slope graphs, all you need to do is add reference lines, which you can do by right-clicking the x-axis and then choosing Add Reference Line. The trick is to make the reference lines equal the minimum month and maximum month. These are dynamic aggregations that will update every time the date range changes. Here's how adding a reference line for the minimum date looks:

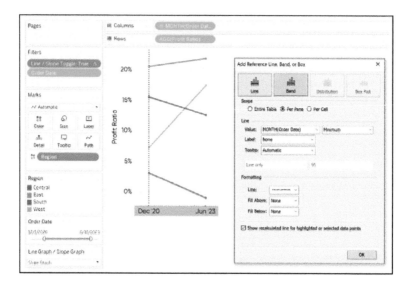

Next, add a second reference line for the maximum date:

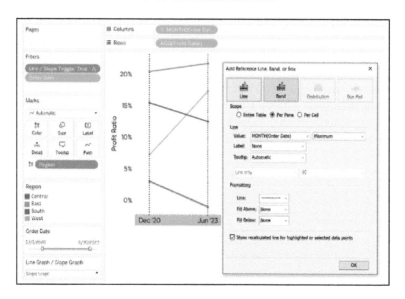

Not only will these vertical lines automatically update based on the date range selected, but reference lines provide more formatting flexibility compared to our approach with discrete months.

Slope graphs are one of my favorite alternatives for visualizing time, and these two techniques for providing a nice aesthetic only make them better!

CHAPTER 28

3 Ways to Make Magnificent Maps: Tip 1

Maps are one of the most effective chart types in Tableau and are also among the easiest chart types to create. They are effective because they help us decode latitude and longitude combinations almost instantly, allowing us to see patterns between geographic locations that might otherwise be challenging to discover. They are easy to create because Tableau comes prepackaged with thousands of geographic coordinates all over the world. This makes it so that simply double-clicking a dimension that Tableau recognizes as geographic will create a map on the view.

What's more, Tableau maps are technically scatter plots with points at the combination of each latitude-longitude pair and an image of a map in the background. This unlocks even more applications including the ability to map anything—even if it's not related to geography. The next three chapters use a map of my top 10 favorite barbecue restaurants to share three ways to take your Tableau maps to the next level. Tips include how to use the MAKEPOINT and MAKELINE functions, create custom geographic territories, and produce a dual-axis map using a combination of generated and custom coordinates.

How to Make Hub-and-Spoke Maps in Tableau

Hub-and-spoke maps not only plot locations on a map, they draw connecting lines between the origin and one or more destinations. We can use this type of map to visualize connecting flight routes, travel schedules, migration patterns—or, in the following case—the path between Playfair Data's office and my 10 favorite barbecue restaurants!

This chapter shows you how to make hub-and-spoke maps in Tableau with the MAKEPOINT and MAKELINE functions (available as of Tableau 2019.2), but I take this a step further and share with you how to use these functions without adding extra rows *or columns* to your dataset, and when *not* to use these new functions.

By the end of this chapter, you will be able to make a hub-and-spoke map connecting any origin to one or more destination points:

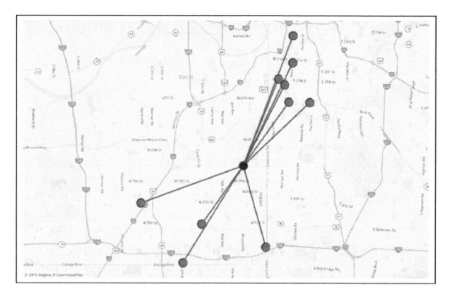

For the next three tutorials, I use the following CSV file that contains the addresses of my 10 favorite barbecue restaurants and Playfair Data's office. You can view and download this dataset from my spreadsheet (*https://oreil.ly/LrkbS*):

	A	B	C	D	E	F	G	H
1	Rank	Location	Address	City	State	Zip Code	Latitude	Longitude
2	1	Gates	1325 Emanuel Cleaver II Blvd	Kansas City	Missouri	64110	39.041166	-94.569253
3	2	Gates	3205 Main St	Kansas City	Missouri	64111	39.068983	-94.585054
4	3	Zarda	11931 W 87th St Pkwy	Lenexa	Kansas	66215	38.97058	-94.725045
5	4	Jack Stack	101 W 22nd St #300	Kansas City	Missouri	64108	39.087733	-94.584767
6	5	Char-Bar	4050 Pennsylvania Ave #150	Kansas City	Missouri	64111	39.053354	-94.592319
7	6	Gates	2001 W 103rd Terrace	Leawood	Kansas	66206	38.939798	-94.60949
8	7	Jack Stack	9520 Metcalf Ave	Overland Park	Kansas	66212	38.955771	-94.668408
9	8	Q39	1000 W 39th St	Kansas City	Missouri	64111	39.057404	-94.598112
10	9	Jack Stack	4747 Wyandotte St	Kansas City	Missouri	64112	39.041225	-94.588637
11	10	Q39	11051 Antioch Rd	Overland Park	Kansas	66210	38.92879	-94.685772
12		Playfair Data	7301 Mission Rd #241	Prairie Village	Kansas	66208	38.996727	-94.630303

Note that my dataset contains columns for Latitude and Longitude; for these examples, we'll use these coordinates instead of the versions that Tableau automatically generates. If Tableau doesn't automatically recognize your coordinates as Latitude and Longitude, you can assign them geographic roles. To do this, right-click each field and then, on the context menu that opens, hover over Geographic Role, and then choose Latitude or Longitude:

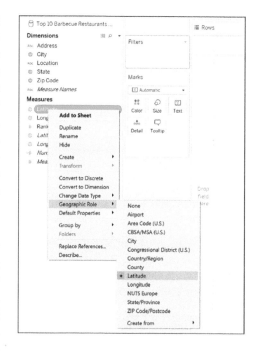

Let's begin with the MAKEPOINT function, which will...well, make points for our origin and destinations. You don't need this function if you are simply creating a symbol map plotting each location at its respective intersection of Latitude and Longitude. However, you can combine making points in this manner with the MAKELINE function in the next step to "connect the dots" between locations with a line.

The syntax for the MAKEPOINT function is MAKEPOINT([Latitude],[Longitude]). We need to use this function in two calculated fields; one for our origin and one for our destination(s).

As you can see in my dataset, I have only one set of Latitude/Longitude coordinates, so let's start with those. This calculated field will eventually make points for my destinations:

```
MAKEPOINT([Latitude],[Longitude])
```

Next, we will need a calculated field that creates the point for our origin. Your dataset might have four columns for geographic coordinates (Origin Latitude, Origin Longitude, Destination Latitude, and Destination Longitude), in which case you would use the Origin coordinates with the MAKEPOINT function to create your Origin points, and your Destination coordinates with the MAKEPOINT function to create your Destination coordinates.

However, as promised in the chapter's introduction, here we look at how to isolate the coordinates for the origin without the need for those two extra columns.

To create the origin point, we use a FIXED Level of Detail (LOD) expression to isolate the coordinates for Playfair Data's office. Here's the formula:

```
MAKEPOINT({AVG(IF [Location] = "Playfair Data" THEN [Latitude] END)},{AVG(IF
[Location] = "Playfair Data" THEN [Longitude] END)})
```

Note that you don't see the expression FIXED in the calculated field. FIXED is the default LOD expression, and when you're not addressing any specific dimension, you don't need to include the word. This is a trick I picked up from my friend Carl Allchin (*https://twitter.com/Datajedininja*) at The Information Lab.

In case it's easier for you to understand what this is doing, you can also write this calculation as follows:

```
MAKEPOINT({FIXED: AVG(IF [Location] = "Playfair Data" THEN [Latitude] END)},{FIXED:
AVG(IF [Location] = "Playfair Data" THEN [Longitude] END)})
```

We are now ready to make our connecting lines using the MAKELINE function. The syntax for the MAKELINE function is MAKELINE([Start Point],[End Point]).

Let's make another calculated field and plug in my [Origin Point] and [Destination Points] calculations:

```
MAKELINE([Origin Point],
[Destination Points])
```

You can also nest the calculations required to create the origin point and destination points, but I find it easier to manage when these are separated into their own calculated fields. If you would prefer to have everything in one calculation, the formula would be as follows:

```
MAKELINE(MAKEPOINT({AVG(IF [Location] = "Playfair Data" THEN [Latitude]
END)},{AVG(IF [Location] = "Playfair Data" THEN [Longitude]
END)}),MAKEPOINT([Latitude],[Longitude]))
```

We're now ready to make the hub-and-spoke map! First, place your Longitude measure on the Columns Shelf and Latitude measure on the Rows Shelf:

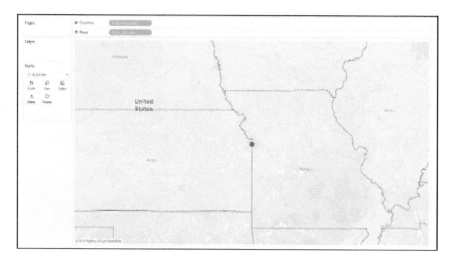

By default, this creates a symbol map with one circle. That one mark represents the intersection of the average Longitude and average Latitude across the entire data source.

When you are creating a map from fields with geographic roles like we are, ensure that you change the mark type from the default, Circle, to Map:

After changing the mark type to Map, simply placing the calculated field created with the MAKELINE function on the Detail Marks Card will draw a line between the origin and each destination:

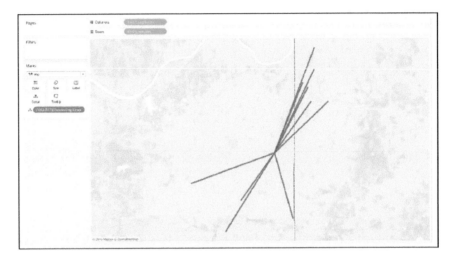

Note you do not always need to use the MAKEPOINT and MAKELINE functions to connect locations with a line. The approach outlined in this chapter is the most efficient way to make a hub-and-spoke map, but if you are simply wanting to connect points in a sequential order, you are better off changing the mark type to Line and adding a field for point order to the Path Marks Card.

 Related: *Practical Tableau*, Chapter 33, "How to Map a Sequential Path" (O'Reilly, 2018)

There are some optional steps if you want to match the hub-and-spoke map pictured in this chapter's introduction, including these:

- Make a dual-axis map with the second axis containing a symbol map. This technique is used to make the larger circle marks at the end of each spoke.
- Change the map style to Streets (available as of Tableau 2019.2). To do this, on the menu bar at the top of the window, click Maps > Map Layers and then change the option selected in the Style drop-down menu on the Map Styles pane.
- Color the marks by whether they are an origin or destination.

- I did not do this in my example, but you can parameterize the origin by simply replacing the hardcoded location (Playfair Data in my example) with a parameter containing allowable values that match each potential origin.

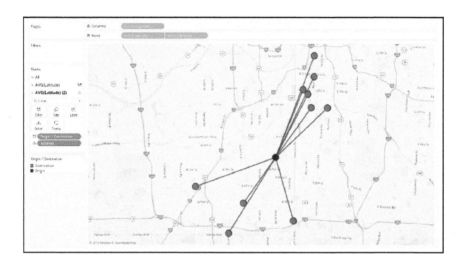

As you can see, Playfair Data is strategically located between my 10 favorite barbecue restaurants in Kansas City!

CHAPTER 29

3 Ways to Make Magnificent Maps: Tip 2

In addition to the features covered in Chapter 7, one of my favorite mapping features added since the release of *Practical Tableau* is the ability to create custom territories. In the Sample – Superstore dataset that comes with Tableau, you will see a dimension called Region that contains custom territories, but that dimension is a field in the underlying data. Now you can create a similar field just by lassoing geographic dimension members on a map view!

How to Make Maps with Custom Territories

For the following illustration, we again use the CSV file containing a list of my 10 favorite barbecue restaurants introduced in Chapter 28. To begin the map this time, I placed the Longitude measure on the Columns Shelf, Latitude measure on the Rows Shelf, and the Zip Code dimension on the Detail Marks Card. I've also changed the mark type to Map to create a filled map showing which zip codes include at least one of my favorite barbecue restaurants and I have filtered out Playfair Data's address:

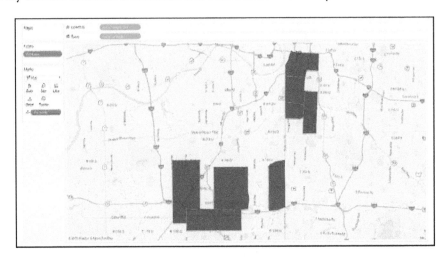

Now, suppose that we want to consolidate the eight zip codes represented into two custom territories: Southwest and Northeast. To begin, access the map controls by hovering over the map. When the controls appear in the upper-left corner of the map, hover over the right-facing arrow and then select the Lasso Selection tool:

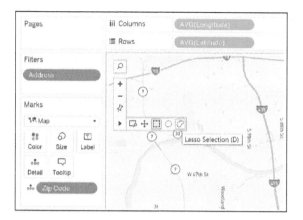

Next, *lasso*—or, trace around—the dimension members that you want to include in the first territory:

After selecting the dimension members, hover over any of the individual dimension members and then, on the panel that opens, click the paperclip icon that appears on the command buttons. This groups the members into a custom territory:

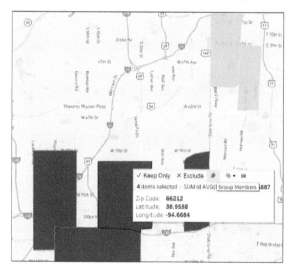

After you group the dimension members, a new dimension is added to the Dimensions area of the Data pane, and that dimension is automatically added to the Color Marks Card. This results in the new custom territory being colored one way, and everything else being colored a second way:

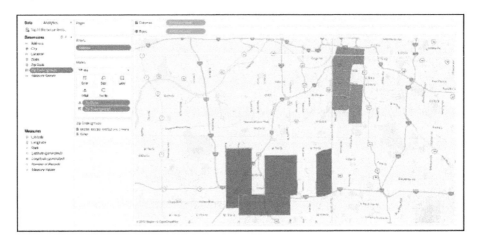

Next, let's repeat the steps to create the Northeast territory:

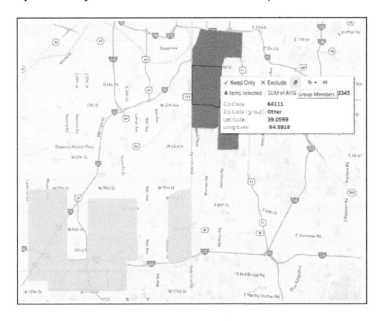

By default, Tableau names these new groupings by the first three dimension members and informs you how many additional dimension members are in the group. It is very easy to modify the names of the custom territories by right-clicking their default name from the color legend and then, on the menu that opens, choose Edit Alias:

You can also rename the new dimension itself by right-clicking it and then choosing Rename.

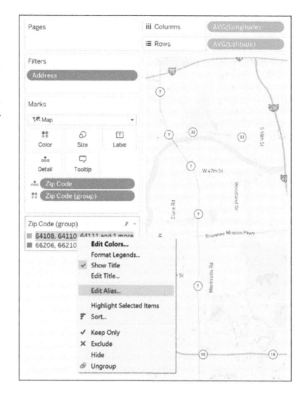

Here's how the final map looks after creating the custom territories, renaming the new dimension, editing the territory aliases, updating the color formatting, removing the Zip Code dimension from Detail (which is no longer needed because the zip codes are part of the custom territory in this case), and adding the Number of Records measure to the Label Marks Card to see how many restaurants are in each territory:

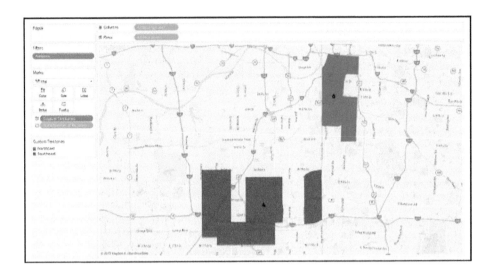

It turns out more of my favorite barbecue restaurants are located in the Northeast territory compared to the Southwest territory.

Another huge benefit of this feature is we now have a new Custom Territories dimension that we can use to create any chart, regardless of whether it is a map. Here's a bar chart showing the average rank per custom territory:

In this case, lowered numbered ranks are better. This bar chart shows me that not only are more of my favorite barbecue restaurants in the Northeast territory, but on average, they have a better rank compared to those in the Southwest territory. An insight made possible by creating custom territories on the fly!

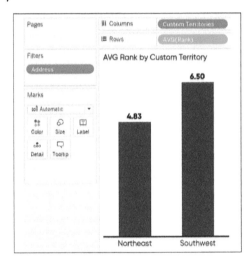

CHAPTER 30

3 Ways to Make Magnificent Maps: Tip 3

Combine Generated and Custom Lat-Long Coordinates

As of Tableau version 2018.1, you can create a dual-axis—or, layered—map, even with a combination of generated and custom latitude/longitude coordinates. This makes it possible to combine specific addresses that require custom latitude and longitude pairs in the underlying data with less granular geographic data that uses the generated coordinates in Tableau.

This has many applications. One of my favorite examples is when I worked with a client that sold its product exclusively online but had competitors that had traditional brick-and-mortar locations. We created a dual-axis map with the client's sales on one axis and the store locations of the client's competitors to identify threats and opportunities.

Let's keep going with this barbecue example to illustrate how this is done. Let's pretend the year is 2036. I have retired from writing about Tableau and got the crazy idea to compete with the greatest barbecue restaurants on the planet by shipping Kansas City barbecue via drone. I have a data file that shows my online orders by city and another file with the restaurant locations from Chapters 28 and 29.

If you are in a similar situation with two separate files, the first thing you must do is a full outer join to bring the dataset with the custom coordinates into the dataset that will use the generated coordinates.

Here's how the data prep piece looks within Tableau Desktop:

To get this to work, the first map must use the generated Latitude and Longitude measures from Tableau. These will be in your dataset automatically if you are using any field that Tableau recognizes as geographic. In my case, this is my online order data. I don't have specific addresses that require custom latitude and longitude coordinates, but Tableau generated coordinates for me because the dataset has the City and State dimensions. Next, we make a symbol map by just double-clicking City and adding Orders to the Size Marks Card:

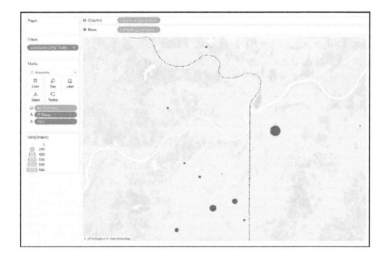

Beware that joining data sources can create duplicate rows. You might have noticed that the aggregation of my Orders measure on the Size Marks Card is Minimum, which takes the smallest number from each City-State pair to deduplicate the rows. For more discussion on deduplicating joined rows, see Chapter 50.

Next, duplicate the map by placing the generated Latitude measure on the Rows Shelf a second time:

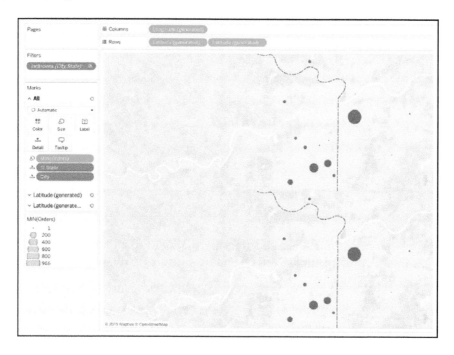

At this point, we have the same map on two rows, but each row has its own Marks Shelf. This means that you can independently edit the maps.

If your custom coordinates are named something other than Latitude and Longitude, you will need to take the extra step of assigning a geographic role to your custom measures. This is accomplished by right-clicking each measure, and then, on the menu that opens, hover over Geographic Role, and then make your selection.

The trick to combining generated and custom geographic coordinates is to place the custom Latitude and Longitude measures on the Detail Marks Card and convert them to dimensions. You can convert measures to dimensions by right-clicking them and then choosing Dimension:

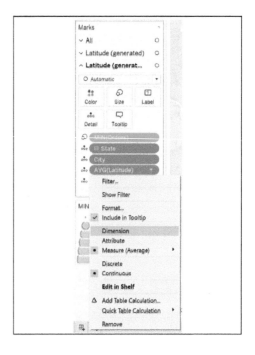

Here's how the maps look after updating the Marks Cards for the second row to include only the custom Latitude and Longitude measures as dimensions:

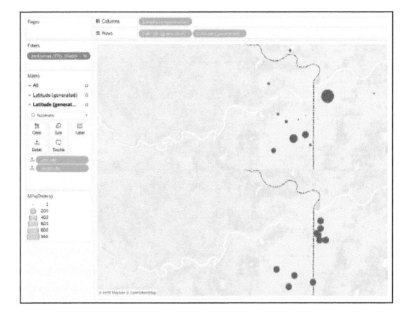

At this point, the top row is showing online sales by city and the second row is showing the physical locations of competitors. What's special is that we are using a combination of generated and custom latitude/longitude pairs!

Let's now combine the layers into a dual-axis map by clicking the second Latitude measure on the Rows Shelf and then selecting Dual Axis:

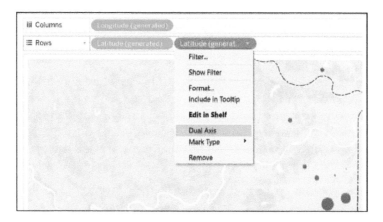

Here's what the final map looks like after making circles the online orders (sized by sales), and black diamonds the competing restaurants:

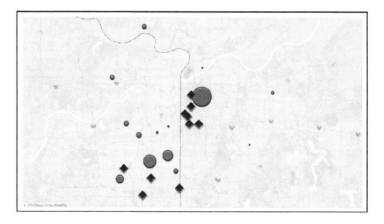

It looks like the biggest opportunity is northwest, where the competitors do not have a presence. Maybe I'll buy a billboard there. We're also doing fairly well down south in the empty space between physical restaurants. Maybe we should advertise in the popular mall that's located nearby. Both are insights and actions that would not have been visible without this dual-axis map using a combination of generated and custom latitude/longitude coordinates!

CHAPTER 31

How to Make a Custom Geographic Map

To me, one of the most undervalued features in Tableau is its ability to visualize maps. Simply double-clicking a field that Tableau recognizes as geographic will generate a symbol map, saving us hours of manual work. As discussed in "Three Ways Psychological Schemas Can Improve Your Data Visualization," Chapter 75 of *Practical Tableau* (O'Reilly, 2018), the spatial context that maps provide help us and our end users to process data more efficiently.

Did you know you can create a map out of literally anything in Tableau? My best known examples of this are stadium maps (*https://oreil.ly/b1XZj*) and showing the source of concussions (as seen in Chapter 11), but the same principles we used to create these maps can also be used to create custom *geographic* maps. Creating custom maps is a relatively easy way to set your reports apart and/or get your maps *in-brand*.

This chapter shows you how to map a background image in Tableau with *any* map image.

How to Map a Background Image to Create a Custom Geographic Map in Tableau

To help illustrate how to create a custom map in Tableau, here I re-create one of the maps featured in my visualization, *Where's Ryan?*. I used the maps to show where I would be speaking in 2017, and I wanted them to be minimalist and represent my brand.

Here's how it turned out:

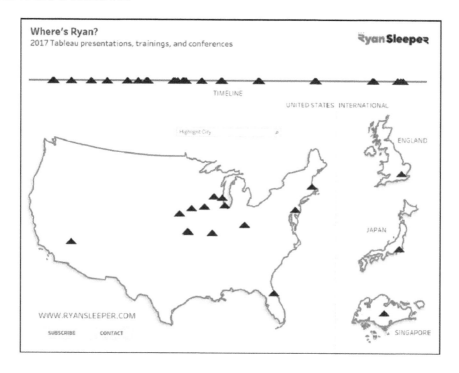

To start the map of the United States, I first needed a silhouette of the country to which I could add my design. There is legal precedence that people can't own silhouettes of objects. Things get grayer if the original artwork is a silhouette, but nobody owns the shape of the United States. For that reason, I was able to simply do a Google search for "US map PNG" to find the foundation for my map.

You can use any image as the foundation, but I recommend using a PNG image type because the background will be transparent. This makes it easier to add effects like strokes and shadows. You can map any image you choose and add the effect however you know best, but I chose to make a simple map with a white overlay, red outline, and subtle shadow. This is not a book on Adobe Photoshop, but for context, here is how my underlying custom map looks:

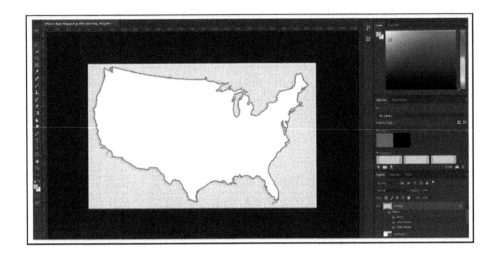

It's important to note the dimensions of the image that you are using. You can find this on a PC by simply hovering over the image in the folder where it is saved. My US map is 1,000 pixels × 619 pixels. We use these dimensions in the next step, which is to create an Excel or Google Sheet (which is what I used) with the fields needed for your visualization, plus a column for the X coordinate, and a column for the Y coordinate.

In the first row, put the width of your custom image as the X coordinate, and the height of your custom image as the Y coordinate.

Here's how my data source looks at this point:

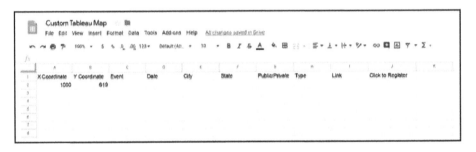

Next, let's start a new Tableau workbook and connect to the Google Sheet or Excel file.

In a new worksheet, click Map > Background Images > [the data source you are using] on the menu bar at the top of the window. This opens the Background Images dialog box in which you can choose your newly created custom background image or map:

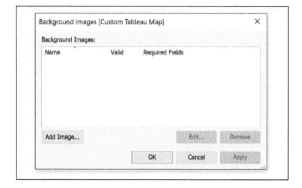

After you click Add Image, browse to find the image that you want to use, set the range of the X Coordinate field to be 0 to the width of the image (1,000 in my case), and set the range of the Y Coordinate field to be 0 to the height of the image (619 in my case):

Place the X Coordinate field from your data source on the Columns Shelf and the Y Coordinate field from your data source on the Rows Shelf. This should show you the background image and one circle, which represents the maximum height and width of the image. Here's how my background image looks with a circle at 1,000 pixels on the x-axis and 619 pixels on the y-axis:

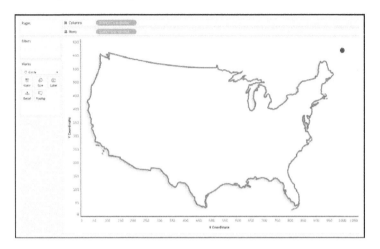

Now that we've laid the foundation, you can look up X/Y pairs to add new marks to the custom background image. The easiest way to do this is add an annotation by right-clicking the view, hovering over Annotate, and then clicking Point:

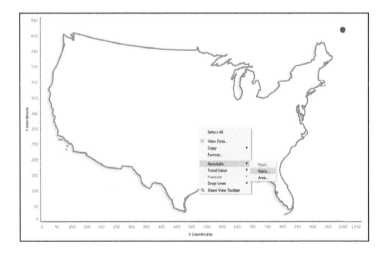

By default, the annotation includes the fields being used on the view. We are using only the X Coordinate and Y Coordinate fields, so those are the fields you will see on the annotation. When the annotation is on the view, you can move the point being annotated by dragging and dropping it anywhere on the image; the annotation auto-matically updates.

Here's how the annotation looks when moving it to Kansas City:

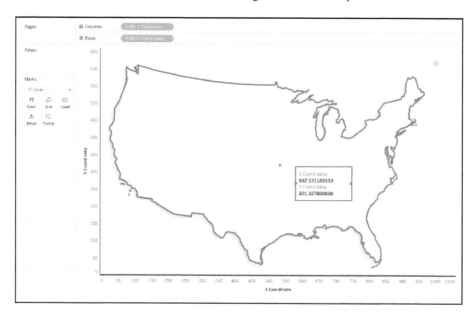

Now, let's go back to the data source to add a record, ensuring that the coordinates for Kansas City are recorded in the respective X Coordinate and Y Coordinate fields. Note that decimals are not needed, so you can round the coordinates:

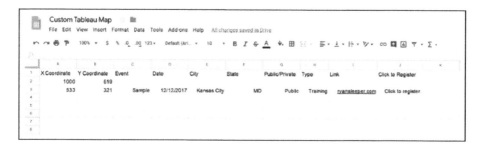

After you refresh the data source and place the Event field on the Detail Marks Card, a new mark will appear on Kansas City. It's important to place a unique identifier for each mark (Event in my case) on the Detail Marks Card; this places a mark at the correct coordinates instead of adding the coordinates for multiple rows together:

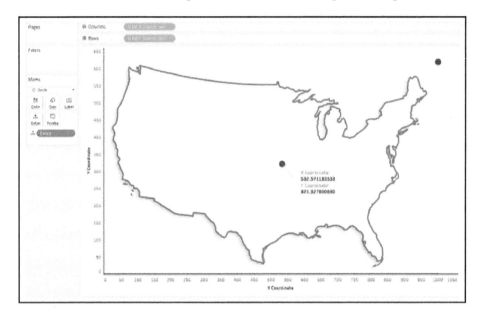

You can repeat these steps to add as many marks as you want to a custom background image. To finalize the view, you can hide the axes, filter out the placeholder mark, and remove the annotation.

CHAPTER 32

How to Make Trellis/Tile/Small Multiple Maps

A chart type with many names—trellis maps, tile maps, or small multiple maps—can help you compare measures across multiple maps in one concise view. No matter what you choose to call them, these maps present an effective way to add context to a view without making your end user do additional work. As discussed in Chapter 25, "How to Make Small Multiples," in *Practical Tableau* (O'Reilly 2018), adding context in this manner is a tactic for avoiding the dreaded question, "So what?"

This chapter provides two techniques for creating trellis/tile/small multiple maps in Tableau. In the first approach, we use table calculations to automatically generate a grid for the maps. I provide the formulas so that you can create these maps in a matter of seconds. In the second approach, we use IF/THEN logic to manually generate the grid. This approach gives you complete control over the number of rows and columns in the layout and in which cell you want each map to appear. I also share a creative hack for how to add a label to each individual map within the view.

How to Make Trellis Maps with Automated Grids

For both approaches to creating small multiple maps in Tableau, we re-create the following view from my Tableau Public visualization: *A Tale of 50 Cities*. In the view, I used a trellis layout to show population changes during each US census dating back to 1790:

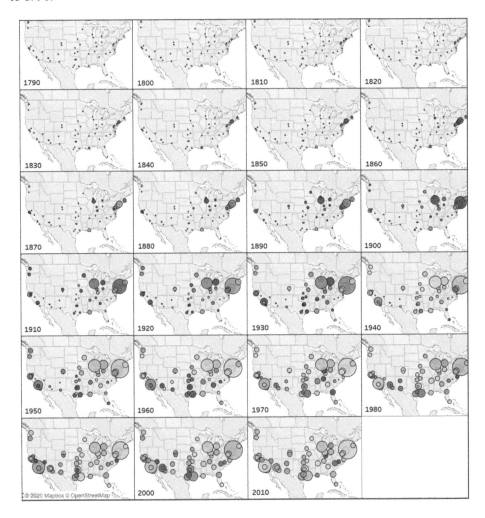

For the first approach—and the approach used to create the preceding view—we use the INDEX() and SIZE() functions to create a dimension for the columns and a dimension for the rows. To my knowledge, this approach was invented by my friend and data viz hero, Chris Love.

First, create a calculated field for the x-axis (or columns); here's the formula:

```
(index()-1)%(int(SQRT(size())))
```

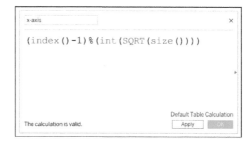

Also create a separate calculated field for the y-axis (or rows):

```
int
((Index()-1)/ (int(sqrt(size()))))
```

Now, let's lay the foundation for the map. Of course, there are infinite possibilities of things you can map, so you will need to apply the following steps to your own visualization. To continue with the example in this chapter, just double-click the City dimension to start the map and place the Census dimension on the Detail Marks Shelf:

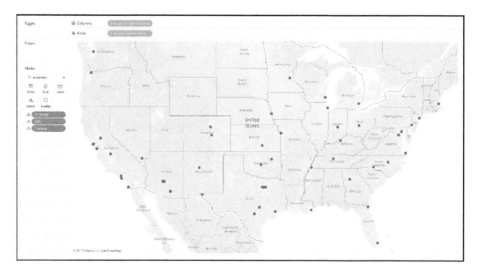

Right now, there are 1,150 marks on the view, but we can see only 50 because they are laying on top of each other (i.e., multiple marks for the same state). Next, we create a grid using the newly created x-axis and y-axis calculated fields, which spreads the marks out into their own respective cells.

First, place the x-axis calculated field on the Columns Shelf, and the y-axis calculated field on the Rows Shelf:

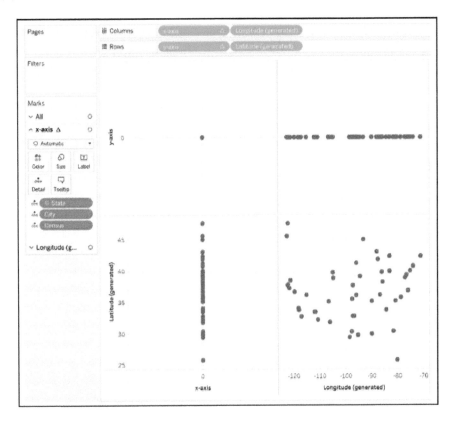

Your keen intuition might be telling you that this doesn't look quite right. That's because we need to change the x-axis and y-axis fields to discrete. To do this, right-click each one (or click the triangle that appears when you hover over the pill) and then, on the menu that opens, choose Discrete:

After changing the newly created calculated fields to discrete, we end up with our original map:

Lastly, we need to change the table calculations on the x-axis and y-axis measures, respectively, so that they compute using the field we want in each cell. In this case, I want a unique cell for each census. To edit the table calculation, you can click each pill again, and then, on the menu, hover over Compute Using, and then choose Census (or whatever field for which you're creating unique cells):

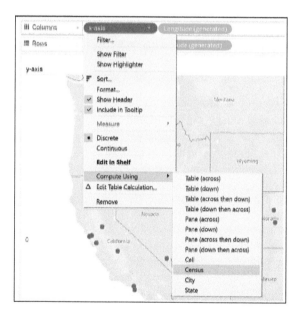

After this change, we have a trellis/tile/small multiple set of maps:

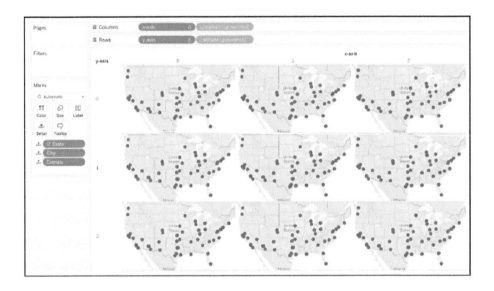

If you follow these steps, but see more than a single dimension member in a cell—or not enough unique cells for the scenario that you're trying to visualize—it is likely that your view includes extra dimensions that must be added to the addressing of both table calculations. This gets trickier, but it's easiest to work with if you right-click each table calculation that is generating the trellis effect (x-axis and y-axis, in my case) and then click Edit Table Calculation.

If you want more cells to be included based on additional dimensions, choose Specific Dimensions and then select the checkboxes for the dimensions that you want included:

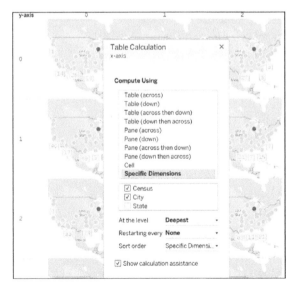

After adding the City dimension to both the x-axis and y-axis fields, the grid has 15 rows by 13 columns, or 195 cells.

You might come across dimension members that contain a varying number of rows. For example, the earliest census data in my dataset contained only four cities: New York, Philadelphia, Boston, and Baltimore. This was problematic because the grid needs to be built to contain the top 50 cities across all 23 census years in my visualization. This mismatching number of census years and city names throws off the dynamic INDEX() calculation that creates our ideal number of columns and rows.

I handled this issue by padding my underlying data with all top 50 cities, even if they did not have data for a certain census. Then, I used dynamic tooltips (Chapter 57) to explain when the city didn't have data. Another way to do data *densification* in this way is to add all relevant dimensions to the addressing, but then click the "At the level" dropdown box and change it from Deepest, back to the dimension for which you want unique cells:

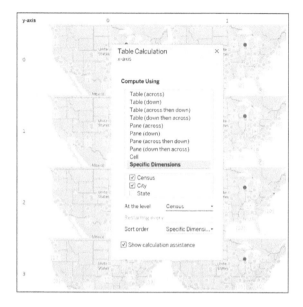

After changing the level of the table calculation to Census, I am back to a grid of six rows by four columns, or 24 cells; one for each census in my dataset (the last cell is blank because I have only 23 census years).

Related: *Practical Tableau*, Chapter 13, "An Introduction to Table Calculations" (O'Reilly, 2018)

Now that you have your maps in a grid, you can modify them as you would like. In my final example, I've added a measure called Population to the Size Marks Card to show the population during each census. I also colored the marks with a table calculation that computed the census over census population change.

How to Make Trellis Maps with Manual Grids

The second approach to creating grids to place your maps in has been covered by my friend and mentor, Ben Jones at DataRemixed (*http://DataRemixed.com*). Although the approach is more manual, it could be the best choice when you're making small grids and want to control exactly which cell each map is created in.

For this technique, as with the first approach, you create a field for the Columns and Rows. The difference is that you specify exactly which column or row the dimension member (in my case, census year) will appear in. Here are my formulas for the Columns and Rows using a subset of my census years:

Column

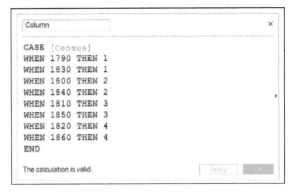

```
Column                                              ×

CASE [Census]
WHEN 1790 THEN 1
WHEN 1830 THEN 1
WHEN 1800 THEN 2
WHEN 1840 THEN 2
WHEN 1810 THEN 3
WHEN 1850 THEN 3
WHEN 1820 THEN 4
WHEN 1860 THEN 4
END

The calculation is valid.              Apply    OK
```

Row

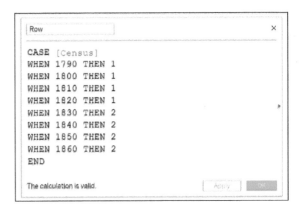

```
Row                                                 ×

CASE [Census]
WHEN 1790 THEN 1
WHEN 1800 THEN 1
WHEN 1810 THEN 1
WHEN 1820 THEN 1
WHEN 1830 THEN 2
WHEN 1840 THEN 2
WHEN 1850 THEN 2
WHEN 1860 THEN 2
END

The calculation is valid.              Apply    OK
```

When used together on the same view, the year 1800, for example, will appear in the second column and the first row. When you use this approach, you want to make sure that there isn't overlap (where more than one dimension member ends up occupying the same cell). This is the reason this approach is better for smaller grids of 12 or fewer cells.

After you've created the two new calculated fields, as we did before, we now add the newly created calculated fields to our foundational map and change the pills from continuous to discrete:

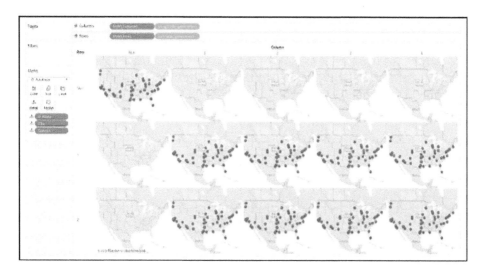

For illustration purposes, I mapped out only eight cells. Because I have more census dimension members than cells, there is a null category that is being placed in the upper-left cell. We can easily filter out this cell by right-clicking one of the Null Headers and then clicking Exclude.

Here's how my final grid foundation looks after filtering out the nulls and hiding the Column/Row headers (right-click them and then deselect Show Header):

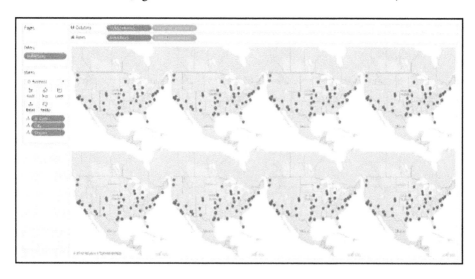

A Creative Hack for Adding a Label for Each Cell in a Grid of Maps

One of the challenges with both approaches to small multiple maps outlined in this chapter is that you can't immediately determine what is represented in each cell. You could float text boxes or add annotations in Tableau to help your users navigate, but as promised, I'm sharing a slightly more efficient and creative way to label the cells.

First, put the field you want used as the cell labels on the Label Marks Card. In my case, I want each cell to display which census is being evaluated, so I drag the Census dimension to the Label Marks Card:

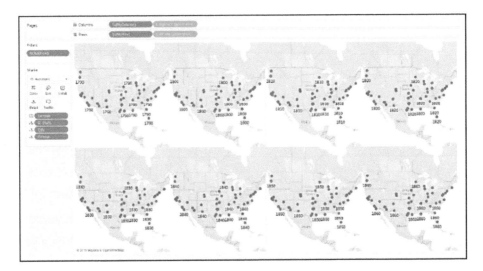

As you can see, the census for each cell is being shown as many times as possible without overlapping. This is redundant and not what we want, so for now, turn all the labels off by clicking into the Label Marks Card again and clearing the Show Mark Labels checkbox.

The trick to getting the cell labels to show up only once per cell but with consistent formatting is to force the mark label to show for only one mark in each cell. To do this, right-click a mark, hover over Mark Label, and then pick Always Show:

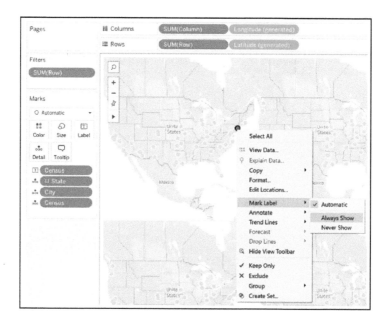

After doing this for each cell, you will have one label per cell. For consistency, I put all of the census years in the lower-left by left-clicking and dragging the label in each cell. This one-time step is a bit manual, but the good news is that, now, if you edit the format of the labels by clicking on the Label Marks Card, all of the labels will update at the same time:

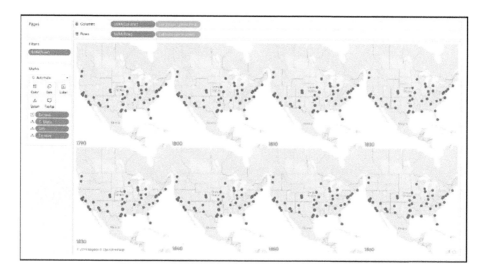

From here, we can use the Marks Cards to encode the symbols by size and color to evaluate how each city's population has changed from census to census. These side-by-side small multiples are packed with comparisons and context that help us avoid causing our audience to ask, "So what?"

CHAPTER 33

How to Make a Timeline

Timelines are not an out-of-the-box chart type in Tableau, but they can serve several practical purposes for your analyses and user experience. First, you can use a timeline in Tableau as a method for showing end users when notable events occurred in the business. For example, you can provide context by lining up a timeline of marketing promotions with a trend line to see when spikes align with your marketing efforts. Second, you can use a timeline as a calendar showing upcoming dates of interest. In both use cases, you can use:

- A relative date filter in Tableau to dynamically display a subset of dates
- Add dashboard actions to link to more information about notable events/dates on the timeline

This chapter shares how to make a timeline in Tableau and how to add an optional reference line to display the current day.

How to Create a Timeline of Key Events in Tableau

To illustrate how to make a timeline in Tableau, we reverse-engineer my *Where's Ryan?* visualization that we use in Chapter 31. In the dashboard, I have a timeline at the top to display the next 90 days of speaking events. The final timeline looks like this:

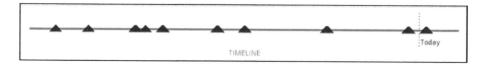

The trick to building a timeline in Tableau is to create a placeholder calculated field. This calculated field will eventually be used to make all the dates line up in a straight line on the timeline. The formula for my placeholder field is simply MIN(0):

For a horizontal timeline like the one at the beginning of this section, place the newly created Placeholder calculated field on the Rows Shelf:

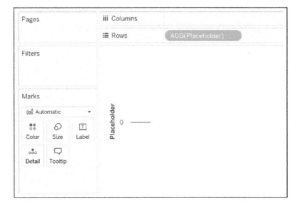

Next, place the Date field that you want to use as your timeline on the Columns Shelf. At this step, I like to choose Date (Continuous) as the date part so that I can use the most granular date available (in this case, day) and use the field as continuous. To do this on a PC, right-click and drag the date field onto the Columns Shelf and pick the very first option that is presented after letting go of the field. At this point, our view looks like this:

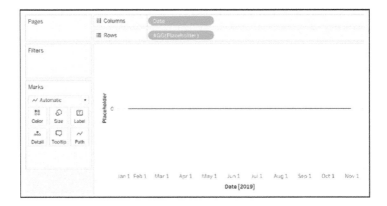

We've now laid the foundation of the timeline. To display events on the timeline, drag your dimension that represents your events to the Detail Marks Card. Note that for the events to line up in the correct place, each event needs a corresponding Date (the same field that is on the Columns Shelf) in the underlying data.

Here's how my timeline looks for each speaking engagement I currently have scheduled for the year (the Event and Link dimensions generate a unique mark for each date):

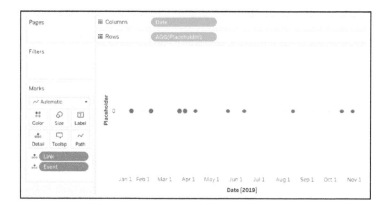

Because some of my dates have passed and most events are not scheduled more than a few months in advance, let's add a relative date filter to show only the next 90 days. To do so, drag the Date dimension to the Filters Shelf and then choose Relative Date. From here, you can set the logic to be any subset of dates that you want to dynamically display:

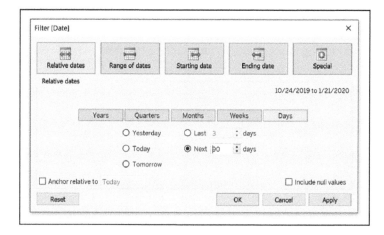

Here's how my view looks after limiting the date range to the next 90 days and changing the mark type to Circle:

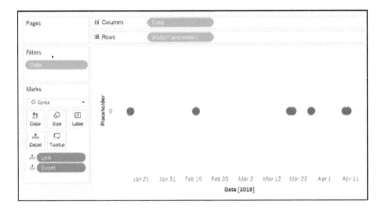

Lastly, a few points on formatting:

- Because the Placeholder field equals 0, all dates will always line up on the y-axis (for a horizontal timeline) of 0. This means that we can format the *zero lines* on the view to give the timeline itself a customized look. You can control the weight, format, and color of the zero lines.

- You should hide the Placeholder header by right-clicking the y-axis and then deselecting Show Header.

- I don't like the light gray lines that still show after hiding the header. These are called axis rulers; to turn them off, right-click anywhere in the view and then, on the menu that opens, choose Format, click to the Lines tab, and choose None for Axis Rulers.

Here's how my timeline looks after a few formatting tweaks:

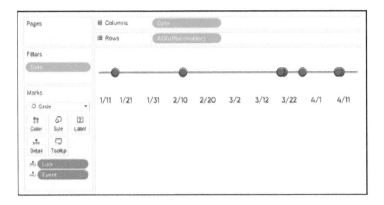

This sheet can now be used alongside other views on a dashboard to provide context or a calendar of events. If you want to link to events on the timeline, use URL dashboard actions.

Related: *Practical Tableau*, Chapter 56, "Three Creative Ways to Use Dashboard Actions" (O'Reilly, 2018)

How to Display a Reference Line for the Current Date

If you would like to take your timeline a step further by dynamically displaying the current date, begin by setting up a calculated field to isolate today's date.

Here's the formula:

    TODAY()

Place this newly created calculated field onto the Detail Marks Card. This allows us to use it as a reference line:

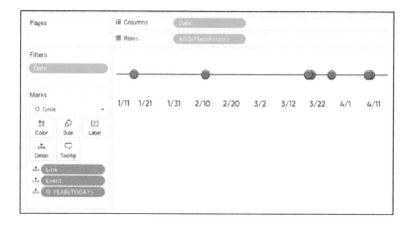

Notice that the default date part for this field after it was placed on the Detail Marks Card is Year (Discrete). The critical step to get this to work properly as a reference line displaying the current date is to right-click the field for today's date and then choose Exact Date:

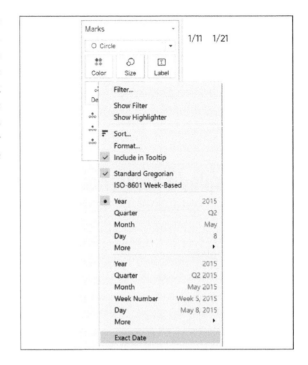

After you've changed this to Exact Date, you can add it as a reference line just as you normally would. For me, I like to right-click the axis, choose Add Reference Line, and then select the Today field as the reference line:

Here's how my final timeline looks after I add the reference line and format it:

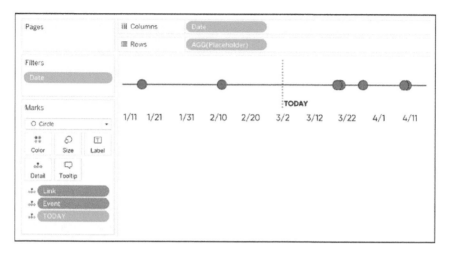

The relative date filter and TODAY() function are dynamic, so this timeline will always update automatically to show a 90-day range and the current date as a reference line!

CHAPTER 34

How to Make a Diverging Bar Chart

A diverging bar chart is a bar chart that has the marks for some dimension members pointing up or right, and the marks for other dimension members pointing in the opposite direction (down or left, respectively). What's unique about a diverging bar chart is that the marks flowing down or left do not necessarily represent negative values. The divergent line *can* represent zero, but we also can use it to simply separate the marks for two dimension members, to represent a goal, or—as is often seen with survey data—to show the break between desired and undesired responses.

The drawback to using diverging bar charts is that it's not as easy to compare the values across dimension members as it is with a grouped bar chart. If, on the other hand, your primary objective is to compare the trend of each individual dimension member, a diverging bar chart is a good option. I also feel that this chart type helps declutter a grouped bar chart, making the data more engaging and easier to understand. In this chapter, we reverse engineer my viz, *50 Years of AFC vs. NFC Matchups* (*https://oreil.ly/YOGs-*), to show you two different approaches to creating diverging bar charts in Tableau.

How to Make Positive Values Diverge Left

Note that if the measure you're visualizing has both positive and negative values for dimension members within the same stacked bar, a diverging bar chart with a divergent line of zero will automatically be created. The following techniques are required when you're trying to create a diverging bar chart with positive values.

The first approach to creating a diverging bar chart in Tableau involves using two different sheets.

For this first example, we re-create a chart like this:

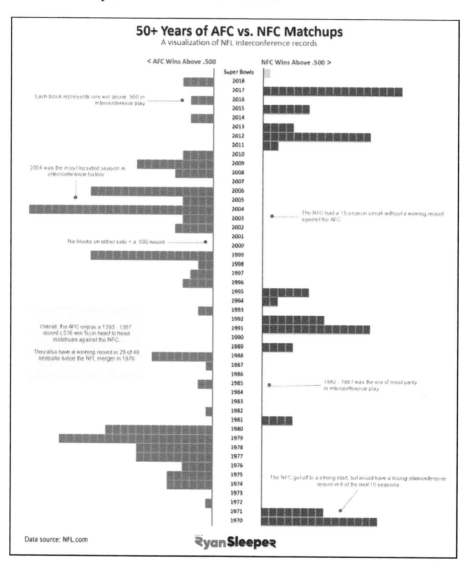

In addition to the clarity provided when trying to view the trend of each individual conference, I thought diverging bars were a good choice for this visualization because the center represents a tie, or a record of .500. This creates kind of tug of war between the dimension members; every block to the left represents one win better than a tie for the AFC and every block to the right represents one win better than a tie for the NFC. This is just one application, but you can also have values that always start at zero and/or stacked bars.

The trick to the first approach is to simply make a bar chart for each side of the diverging bar chart and reverse the axis scale for one of the charts so that the bars point left or down (based on whether your measure is on the Rows Shelf or Columns Shelf).

In this example, the right side of the chart is the primary view and the left side is a second view with the axis reversed and the year numbers hidden (because they are already represented in the first chart). You can hide the header for the second sheet by right-clicking it and then, on the menu that opens, deselect Show Header. I then lined up the two sheets on a dash-board:

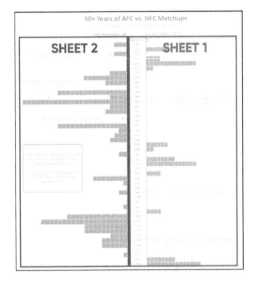

To reverse an axis, right-click it and then choose Edit Axis. This opens the Edit Axis dialog box in which you can choose to reverse the scale:

Making a Diverging Bar Chart on a Single Sheet

It's often best practice to consolidate sheets when possible, which will give you a slight bump in efficiency but also make your workbook easier to manage.

For this second approach, we create a new diverging bar chart using a calculated field. To begin, let's make a stacked bar chart showing wins per conference per season:

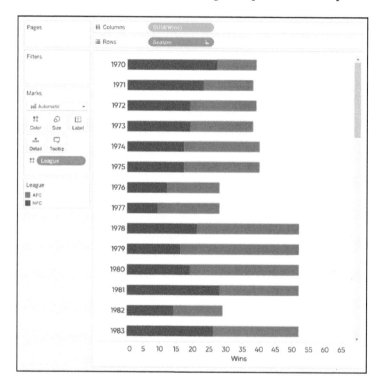

The trick to creating a diverging bar chart on a single sheet is to create a calculated field that multiplies the values for some of the dimension members by negative one. Here's how my calculated field looks to make the values for the AFC go to the left:

```
IF [League] = "AFC" THEN -[Wins]
ELSE [Wins]
END
```

Note that I'm using only two dimension members in this example so I'm making only one go left, but you could make the marks go left for as many dimension members as you want through the use of OR statements. For example, if I were using the Sample – Superstore dataset to make the Sales values for the East and Central dimension members of the Region dimension go left, my calculated field would be as follows:

```
IF [Region] = "East" OR [Region] = "Central" THEN -[Sales]
ELSE [Sales]
END
```

After I replace the original Wins measure with my newly created Diverging Wins calculated field, the chart looks like this:

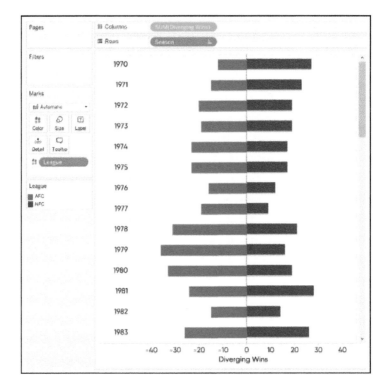

One word of caution with this chart type that I hope you find obvious: the left side of this chart implies that the values are negative. If you are creating a diverging bar chart with positive values on both sides, I encourage you to at least hide the axis and consider adding the positive values to the Label Marks Card:

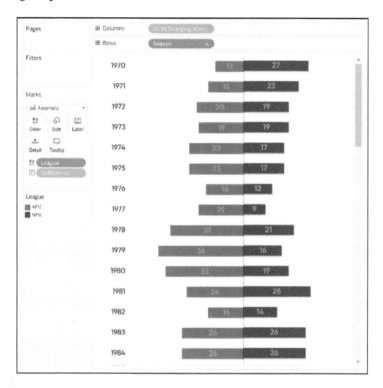

As you can see, one drawback with consolidating both sides of the diverging bar chart into a single sheet is that the years are listed down the left side instead of the center. There is a hack to improve this that involves creating a placeholder calculated field:

After you have the placeholder calculated field, place it on the same shelf as your existing measure:

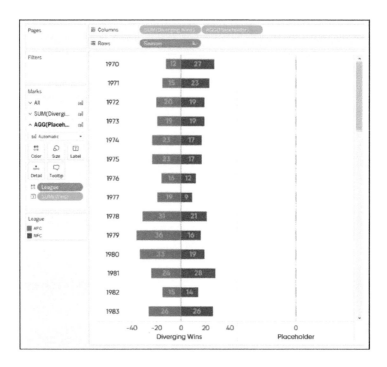

Turn the chart into a dual-axis chart by clicking the Placeholder pill and then choosing Dual Axis.

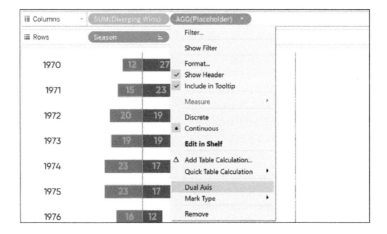

By default, this changes my mark type to Circle, but we'll discuss that in a moment. The most important thing to get the Placeholder measure to line up in the center is to synchronize the axes. To accomplish this, right-click either axis and choose Synchronize Axis:

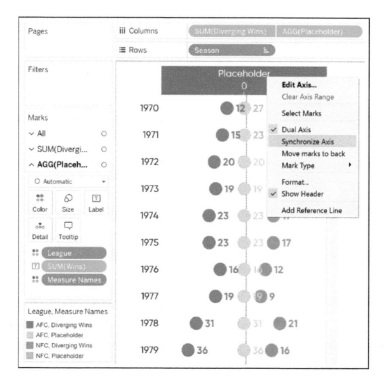

What's important about this is now that there are two measures on the Columns Shelf, they each get their own set of Marks Cards that you can independently edit. This means that I can change the color and mark type of the bars back to the way they were and change the mark type for the Placeholder measure to Text.

Here's how my final view looks after making these changes, hiding all the headers, and adding the Season (Year) dimension to the Text Marks Card for the Placeholder measure:

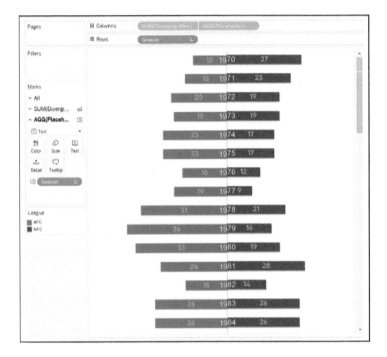

Again, this is just one example for illustration purposes, but there are several possible applications for the second axis. One common alternative is to create a dual-axis combination chart like we've done, but change the mark type for the second axis to Circle and add an informative label. In my case, I could have used the circle to display which conference won that season and/or by how much. You also do not need to lock the primary or secondary axes at zero; unlocking even more possibilities.

CHAPTER 35

How to Make Marginal Histograms

As I often discuss, one of my main objectives as a data visualization practitioner is to avoid causing my audience to ask the question, "So what?" I never want to have an analytics partner put their trust in me and spend my time building out a dashboard, only to have them not know why my findings are important or what to do about them. Three of the best ways to make your data visualization deliverables useful are to build in comparisons, add context, and visualize the same fields in different ways.

In the case of scatter plots, adding *marginal histograms* accomplishes all three of these techniques. Marginal histograms are histograms that are incorporated into the margin of each axis of a scatter plot for analyzing the distribution of each measure. And as my friend Steve Wexler of Data Revelations points out, "They're not just for scatter plots" (*https://oreil.ly/dgIsP*).

This chapter shows you how to make marginal histograms for scatter plots, marginal bar charts for highlight tables, and explains the difference between the two.

How to Add Histograms to the Margins of Scatter Plots

The traditional use of marginal histograms involves adding a histogram of each measure used in a scatter plot to the top and right margin.

To begin, consider the following scatter plot looking at the Profit and Age of Customer (in days they've been a customer) measures by the Customer Name dimension:

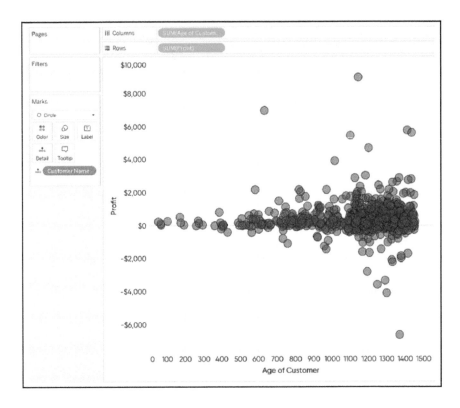

If you want to follow along using the Sample – Superstore dataset, Age of Customer is a calculated field using this formula:

```
DATE(TODAY())-{FIXED [Customer Name]: MIN([Order Date])}
```

This formula takes today's date minus the lowest date (i.e., the date they became a customer) for each customer name.

There are quite a few insights provided by the preceding scatter plot. It seems that there is a correlation between the age of the customer and their lifetime value, which makes sense. As a customer has been retained longer, they've had more opportunities to buy something else and add to their running total of profit. It also looks like most of our customers are bunched between 1,100 and 1,500 days old, and that most of our outlier customers (i.e., extra high or low profit) are at least a year old.

All of this is great, but it would be easier to see the distributions of our two measures if we added marginal histograms to this scatter plot. To start a marginal histogram, create a histogram for each measure on a separate worksheet. Histograms, which are made with just one measure, are one of the few chart types I prefer to make using Show Me. To make a histogram, simply create a new sheet, click the measure from which you want to create the histogram, click Show Me in the upper-right corner of the Authoring interface, and then choose Histogram.

Here's my default histogram for the Profit measure:

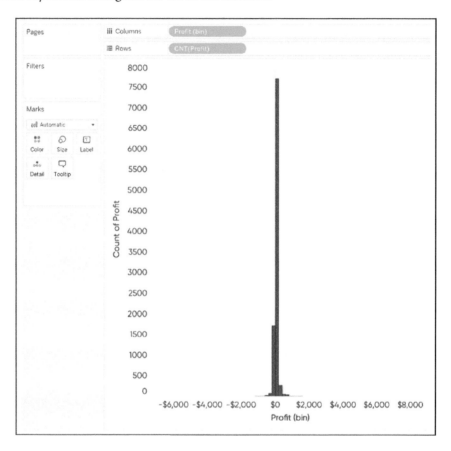

There are two changes that I want to make to this histogram before combining it with the scatter plot. First, because this histogram will eventually reside in the right margin of the view, I want to swap the measures on the Rows Shelf and Columns Shelf, which will give the histogram a vertical orientation. Second, by default, Tableau is generating a *bin* size of 283. I normally like to change this to a more round, human-friendly number, such as 500.

To edit the bin size, in the Dimensions area of the Data pane, right-click the Profit (bin) dimension and then, on the menu that opens, click Edit, and then enter a number in the *size of bins* text box. Here's how the Profit histogram looks at this point:

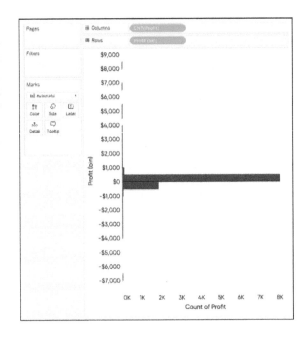

Next, create a horizontal histogram for your second measure. Here's how the histogram looks for Age of Customer (Days) with a bin size of 50:

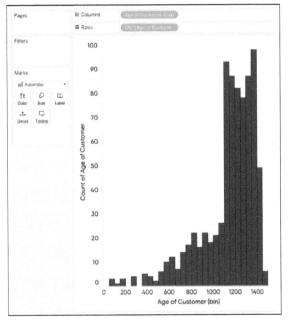

The trick to converting these three separate worksheets into a marginal histograms view is to line them up on a dashboard. You can easily accomplish this in Tableau with tiled dashboard sheets and the Blank object. First, let's add all three elements to the dashboard:

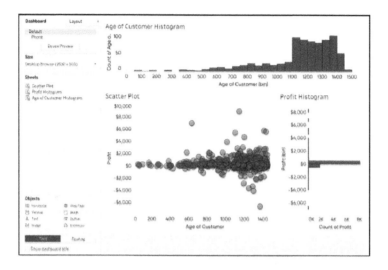

The next step is optional, but I like to clean up the marginal histograms by removing sheet titles and axis headers. Instead of providing the values via the axes, you can add labels directly to the marks:

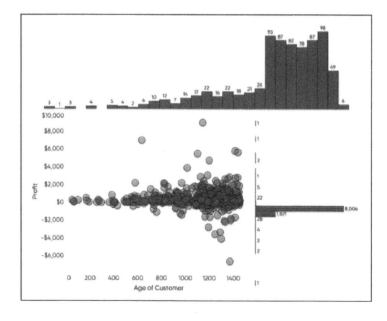

Lastly, add Blank dashboard objects to the left and right of the top marginal histogram and below the right marginal histogram to get them to line up perfectly with the axes of the scatter plot:

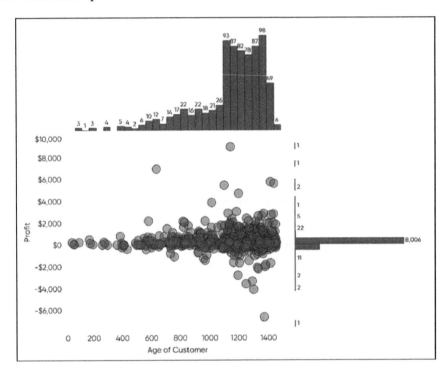

As suspected by looking at the scatter plot view alone, most of our customers are between 1,100 and 1,400 days old, but now we can see the precise distribution. On the y-axis, we're now able to see that most profit values are between 0 and 500, but that our second biggest distribution is actually negative; an insight that would be very challenging to discover with the scatter plot alone.

How to Make Marginal Bar Charts in Tableau

The concept of marginal histograms works well any time you have a breakdown on both the y-axis and x-axis and you want to visualize a higher aggregation than the view's level of detail. To give you an alternate example, consider this highlight table looking at sales by region and segment:

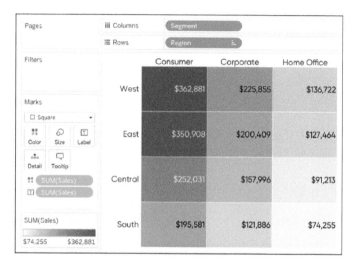

This view does a great job of showing our highest-selling combinations of region and segment, but there is also value in knowing the sales across regions and segments individually (i.e., not combined). To get this latter piece of context, we can create a view similar to the marginal histograms example with the previous scatter plot.

The difference is that the charts in the margins this time will technically be *bar charts* (sales by region and sales by segment). They're not histograms because histograms are made with exactly one measure; that one measure is then broken into counts of the values across equally sized bins. Our bar charts will use one measure (Sales), but also one dimension in each case (Region in the first; Segment in the second).

Here is how a marginal bar chart view looks when added to the preceding highlight table. For best results, sort both bar charts in descending order and ensure that the dimension members of the highlight tables line up properly with the dimension members within the bar charts:

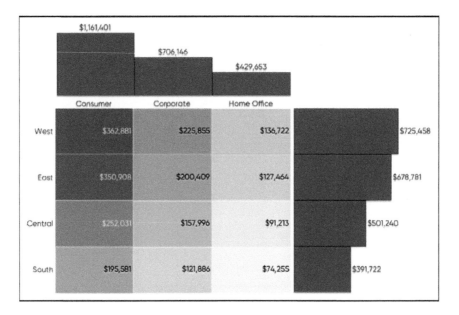

Now, in addition to seeing the highest-selling combination of region and segment, we can view the sales at the region level individually (without a breakdown by segment) and at the segment level individually (without a breakdown by region). This is a more visual alternative to adding subtotals to a highlight table!

CHAPTER 36

How to Make Unit Histograms/ Wilkinson Dot Plots

Unit histograms, or Wilkinson dot plots, show distributions of individual data points instead of bucketing them into bins like traditional histograms. My friend and Tableau Zen Master Hall of Famer, Steve Wexler, suggested this as an alternative to one of my visualizations, and I liked the idea so much that I wanted to show you how to build it in Tableau.

I show how to make one-dimensional unit charts in Tableau in Chapter 62 of *Practical Tableau* (O'Reilly, 2018), which are similar, but unit histograms are slightly more flexible because you can use a mark type other than bar (i.e., circle or shape). By the end of this chapter, you will be able to visualize a distribution of individual items which is effective and engaging. I also show you how to change the mark type so that the distribution looks like little men standing on top of one another's heads which is, well, just for fun.

How to Stack Marks to Create Unit Histograms in Tableau

By the end of this chapter, you will be able to make a version of a dot plot that looks like this:

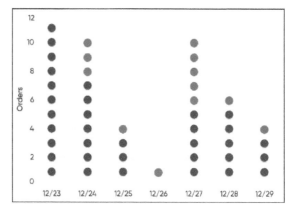

To create this type of chart in Tableau, your dataset must have the dimension members that make up the units on individual rows. To illustrate, I use the Order ID dimension from the Sample – Superstore dataset because every order has an individual row in the underlying data.

If you want to display a vertical distribution, like a traditional histogram, place the dimension you are using on the Rows Shelf with an aggregation of CNTD (count distinct):

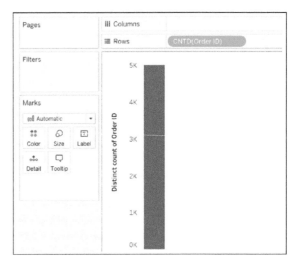

Next, place the dimension that you want to compare on the Columns Shelf. This type of chart is technically meant to compare continuous variables, but in Tableau, you can use the following technique to compare the number of individual units across categorical dimensions. For this example, let's use continuous day of date to see how many orders I have per day:

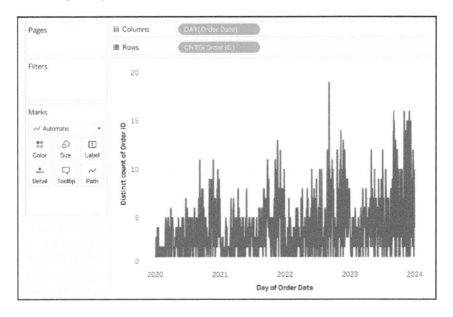

The default mark type for this combination of fields is Line, so the next thing that we want to do is change the mark type to Bar:

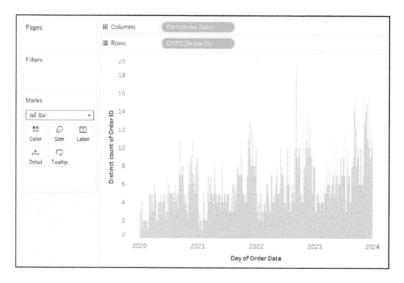

The sample dataset in Tableau has four years of data, which is making our orders by day bar chart very crowded from left to right. To show you the individual days better, let's place a date filter on the Filters Shelf that keeps only the past seven days in the file:

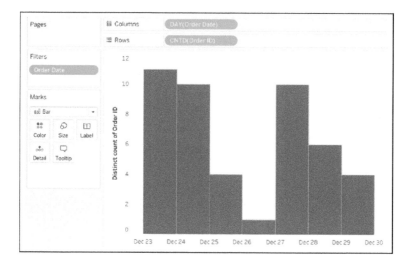

At this point, we technically have a histogram because we are comparing two continuous variables. To see the individual units, place the dimension that you are counting from the Rows Shelf (Order ID in my example) onto the Detail Marks Card:

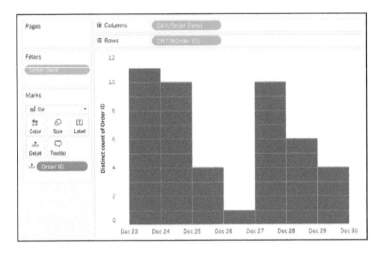

We now have a unit chart like the one at the beginning of this chapter. This is why it is so important that the dimension you are using has individual rows for each dimension member in the underlying data. When that is the case and we place the dimension on the Detail Marks Card, it creates *equally* sized units in our charts. Each block at this point represents an individual order.

For a better look, in my opinion—and to get this to look like a Wilkinson Dot Plot—change the mark type from Bar to Circle:

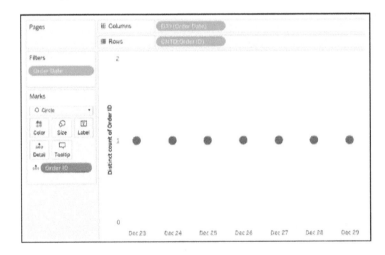

By default, the marks for the Circle mark type are not stacked. So, even though we still have multiple marks in most of our columns, we cannot see the individual units because they are all overlapping on the baseline.

This is the key to creating this type of chart; we must stack the circles atop one another. On the menu bar at the top of the window, click Analysis > Stack Marks and then choose On:

When the marks are stacked, we end up with a unit histogram:

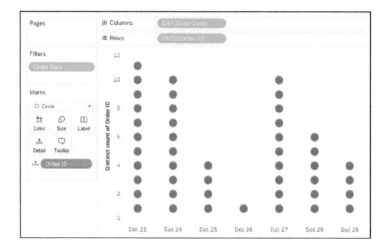

Charts like these that display individual units have been criticized by the likes of Stephen Few because they cause the end user to slow down and count the individual units. I, however, like the added value that comes in Tableau by adding context via the tooltips:

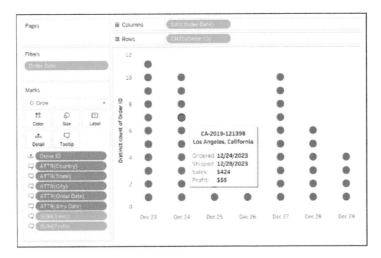

You can also encode the individual marks by color to provide additional context. Here's how my unit histogram looks after coloring the marks by the Profit Ratio measure:

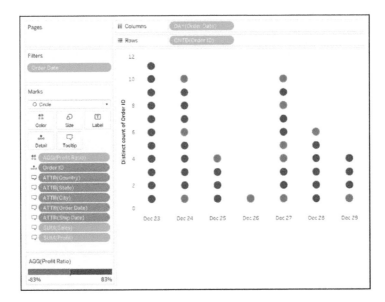

If you want to sort these units so that the profitable and unprofitable orders are grouped together, on the Detail Marks Card, right-click the dimension and then choose Sort:

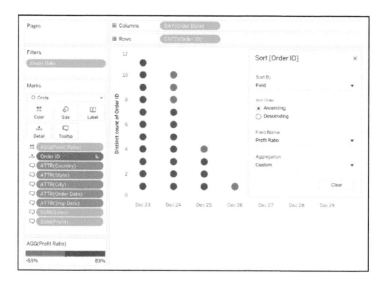

Here's how my final Wilkinson dot plot/unit histogram looks after I format it:

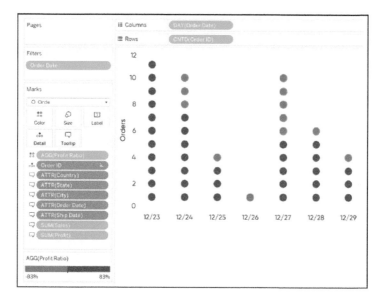

From here, we can use this chart with anything to represent the individual marks! As promised, here's how the view looks if I change the mark type to Shape and then choose a person icon from one of the custom shape palettes:

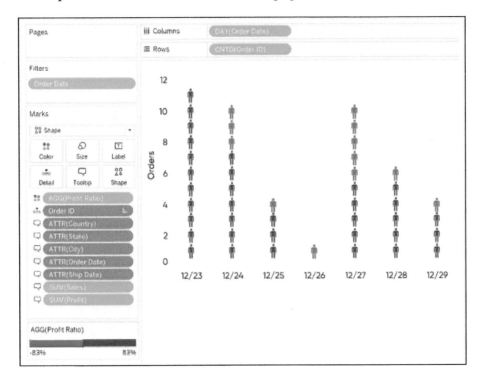

To learn how to integrate your own custom shapes, see Chapter 95.

CHAPTER 37

How to Make a Dual-Axis Waterfall Chart

In this tutorial, we use my *2½ Minutes to Midnight* visualization (*https://oreil.ly/ WkRTb*) to create a waterfall chart in Tableau. A waterfall chart shows the progression toward a cumulative result by showing how positive or negative values contribute to the total. In my data visualization, the waterfall chart was an effective choice for showing how we've moved closer to and further from midnight on the Doomsday Clock since its inception in 1947. We eventually end up at—you guessed it—2½ minutes to midnight. In a corporate environment, waterfall charts can be a great choice for showing how specific segments are contributing to your end goals and/or the makeup of the result.

In addition to the foundational waterfall chart, I'll show you how to use a dual axis to add value to this type of data visualization. In my example, the dual axis is used to display the absolute number of minutes to midnight after each change. In a corporate setting, we can use the second axis to show absolute changes, percent changes, or some other metric of choice. I also like that the dual axis creates a kind of teardrop effect that helps communicate the direction of the change.

How to Use Running Total to Create a Waterfall Chart in Tableau

Before we begin, let's look at how the final visual will look:

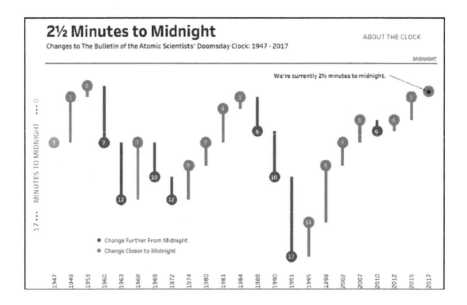

The data used to create this waterfall chart is quite simple, consisting of just four fields: Year, Minutes to midnight, Change, and Reason (not pictured):

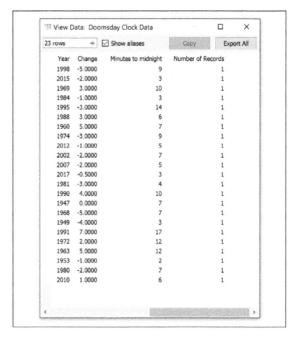

To create a waterfall chart in Tableau, start by placing the discrete dimension that will be used as your breakdown onto the Columns Shelf; on the Rows Shelf, place the SUM of the measure that you are evaluating. I'm looking at *Minutes to midnight* as my measure and Year is the breakdown (Year is being used as a discrete dimension):

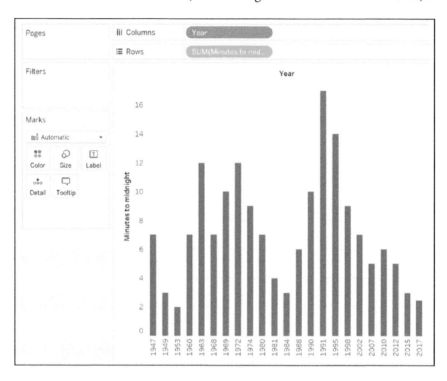

At this point, you have a standard bar chart showing your measure broken down by your dimension. The next step in creating a waterfall chart in Tableau is to change the mark type from Automatic, which is currently Bar, to Gantt Bar. To accomplish this, click the mark type drop-down box on the Marks Shelf and then make your selection.

Note that if you are following along using your own data, my *Minutes to midnight* measure is essentially a running total. If your data field is absolute and not a running total, you can convert it to a running total at this point by adding a table calculation to the measure (click the measure on the Rows Shelf, hover over Quick Table Calculation and then click Running Total).

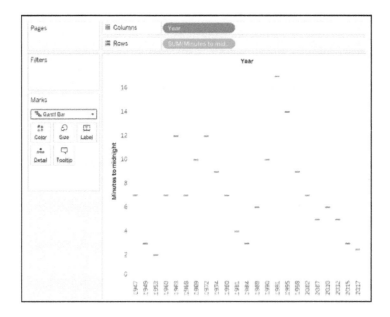

All this did at this point was change each bar into a dash mark at the height of each bar. To create the blocks that will eventually end up as our waterfall chart, place the Change measure (or the absolute version of whatever measure you are using) onto the Size Marks Card.

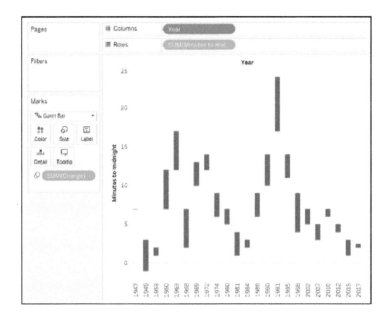

We are getting closer, but our blocks are not lining up as expected. With a waterfall chart, one bar is supposed to start where the bar before it ended. Here is where the magic happens. To get the bars to line up properly, double-click the measure pill that is on the Size Marks Card and type a negative sign (–) before the aggregation:

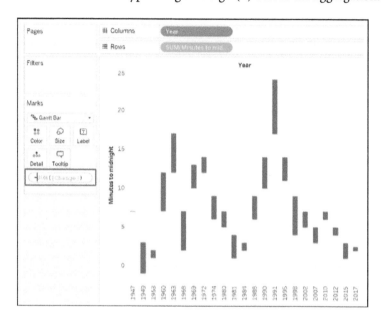

Press Enter to apply the change; the bars then properly line up:

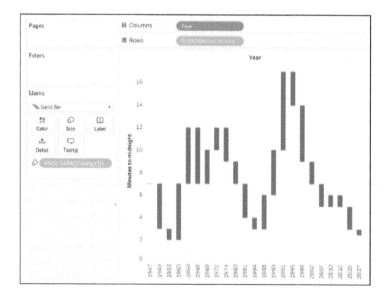

Now we have a serviceable waterfall chart. I have created this visualization with a dataset that was prepared for this specific purpose. If you used a table calculation to create a running total, you have the option to add a grand total to the view by clicking Analysis on the menu bar, hovering over Totals, and then clicking Show Row Grand Totals.

Although this waterfall chart is showing me the progress to and from *midnight*, I thought it would be better to put midnight, or 0, at the top. It's easy to reverse an axis in Tableau: right-click it, choose Edit Axis, and then select the Reversed checkbox:

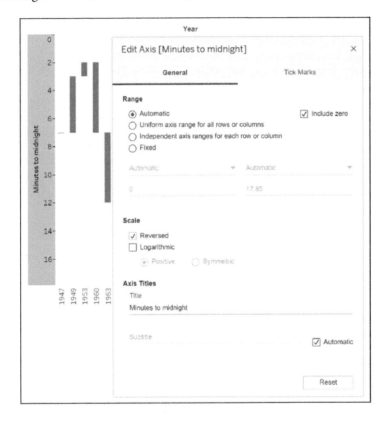

After reversing the axis, you can really begin to see the final shape coming into focus:

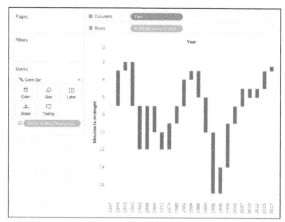

To make this a dual-axis waterfall chart, start by placing the measure of the analysis on the Rows Shelf a second time. To do this, either drag it from the Measures Shelf or hold the Control key (on PC) while you left-click and drag the pill to create a duplicate.

After you reverse the axis for the second row like we did in the previous step, you end up with the exact same chart on two rows:

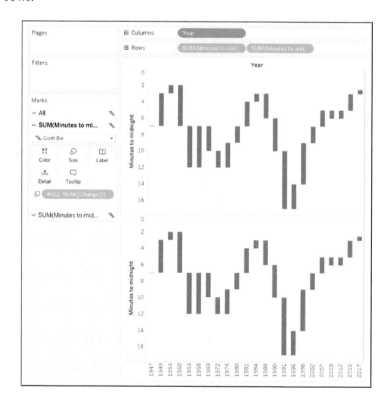

For the second row, let's change the mark type to Circle by going to the Marks Shelf for the second row and changing the mark type from Gantt Bar to Circle. Let's also remove the measure on the Size Marks Card for the second row, which makes all of the circles the same size:

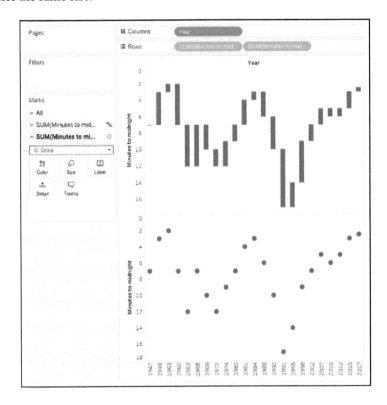

To make this dot plot and Gantt chart share an axis, click the second occurrence of the measure on the Rows Shelf and then choose Dual Axis:

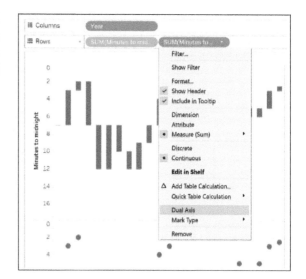

To ensure that the dots line up with the ends of the Gantt bars, right-click the second axis and choose Synchronize Axis:

From here, the rest is just formatting. Here are some additional steps I took to create the final view:

- Hid the second axis by right-clicking it and deselecting Show Header.
- Made the Gantt bars *skinnier* by navigating to the Marks Shelf for the Gantt marks, clicking into the Size Marks Card, and dragging the slider to the left.
- Colored all of the marks by their change (gray for no change; blue for one direction; red for the other direction).

- Added the *Minutes to midnight* measure to the label for the circle marks.
- Added a reference line for *midnight*, an annotation for the current status, and changed the format of the text on the view.

My final dual-axis waterfall chart looks like this:

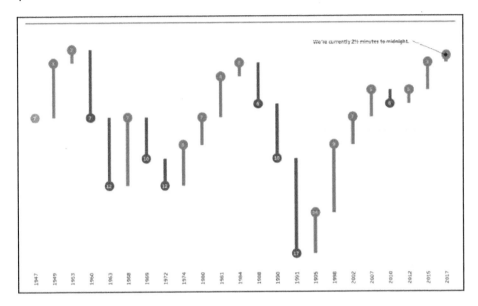

The Gantt marks on the primary axis are creating a traditional waterfall chart showing how performance changed over time and where we ended up. The circle marks on the dual axis help imply the direction of each move and have plenty of real estate to display the performance on each label.

CHAPTER 38

How to Make Performance Indicator Titles

One of my favorite explanatory tactics in Tableau is to conditionally color a dashboard object based on whether the performance improved or declined. This is an explanatory tactic because it provides only a very high-level explanation of how something is performing, but it works well for several reasons, such as it:

- Helps avoid the question "So what?" by providing at least one comparison point
- Provides an alert that a deeper analysis should take place
- Is almost universally understood by end users

We can use these indicator titles in two ways. First, we can use them behind a callout number to give a bold visual cue that the Key Performance Indicator (KPI) featured is performing better or worse (which we discuss in Chapter 39). Second, we can use them without numbers, making the title of a chart act as a basic data visualization itself. This chapter shows you how to create an indicator title as seen in my visualization, *My US Stock Portfolio* (*https://oreil.ly/kuUBY*).

How to Directly Communicate Performance in Chart Titles

Let's begin by looking at the *My US Stock Portfolio* dashboard. As you can see, each widget features a title with the name of the asset.

The titles are colored blue if I've made money since purchasing the stock, and red if I lost money on the investment:

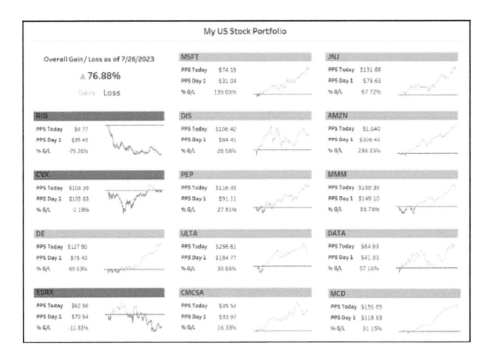

In this case, my comparison formula was the current price of the stock divided by the price I purchased it for. However, this concept can be tailored to fit any comparison point of interest (i.e., previous month, previous year). To help illustrate, and so you can follow along with the same dataset, let's create a similar dashboard title using the Sales measure in the Sample – Superstore dataset.

To begin, we create a comparison. Again, this can be any comparison that is relevant for your business, but let's set up a calculation that compares sales for a reporting month (controlled by a parameter) to sales for a comparison month (controlled by a parameter):

Now that you have the comparison that will eventually control the color of the indicator title, setting it up is easy!

On a new worksheet, change the mark type to Square and place the comparison calculation on the Color Marks Card. Note that my current parameter selections are December 2023 as the reporting month, and November 2023 as the comparison month:

Perhaps the most critical step of this setup is to edit the colors so that they have a diverging color palette, 2 steps, and a center of 0. This ensures any positive change will be colored differently than any negative change. You can edit colors in this way by double-clicking the color legend:

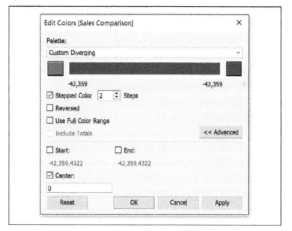

To finalize the indicator title, put the comparison field you are using on the Label Marks Card. After a field is on the Label Marks Card, you can edit it to show anything you want. For this example, let's customize the label to show only the name of the KPI:

Now, when the reporting month underperforms the comparison month, the indicator title turns red; when the reporting month outperforms the comparison month, the indicator title turns blue:

You can add these indicator titles to another dashboard object such as a trendline or bar chart, showing how dimension members ranked for the KPI. In the case of *My US Stock Portfolio*, I placed an indicator title for each stock in a layout container with a text table and line graph. Of course, you can also make the indicator titles a bigger focus by placing onto the Label Marks Card callout numbers, larger text, and/or up and down arrows.

CHAPTER 39

How to Create Performance Indicator Titles with Comparison Score

One of the primary elements that you'll find on almost every high-level corporate dashboard that I ever create is a callout that conveys the current performance of the KPI by some comparison point (i.e., goal, prior period). In Chapter 38, we explored how to make *performance indicator titles* in Tableau, but I usually take these a step further by displaying the KPI's performance and coloring it based on a 100-point index or percent change. I like this element for several reasons:

- It provides the exact numbers for my stakeholders that are accustomed to viewing reports as raw data

- It forces them to benefit from the preattentive attribute of color

- It is explanatory and helps the user decide whether they should invest further time for investigation

This chapter shows you two design options for creating a current versus comparison index callout in Tableau.

How to Create a Current Versus Comparison Index Callout in Tableau

I've been using the first approach to creating a current versus comparison index for years, so let's begin there. The element consists of three rows, each of which will require their own calculated field:

- The current period performance with a mark type of Text

- The goal or prior period performance with a mark type of Text

- The current versus comparison index with a mark type of Square

For this scenario, let's imagine the current performance is sales from the Sample – Superstore dataset during the month of July and the comparison is sales during the month of June. There are several ways to isolate the current performance and the comparison performance. My favorite approach to comparing period-over-period performance is outlined in Chapter 53, but for the purposes of this illustration, I simplify the two calculated fields, as follows:

Current Performance:

```
SUM(IF MONTH([Order Date]) = 7 THEN [Sales] END)
```

Comparison Performance:

```
SUM(IF MONTH([Order Date]) = 6 THEN [Sales] END)
```

So far, the first formula shows sales when the month of Order Date equals 7 (i.e., July), and the second formula shows sales when the month of Order Date equals 6 (i.e., June). The final formula needed is to compute a 100-point index. I love to use a 100-point index score because it normalizes data on a 100-point scale, even with different data formats (e.g., currency, integers, percentages).

It also makes the math easy. A score greater than 100 means that the current performance outperformed the comparison; a score less than 100 means the current performance underperformed the comparison. Because everything is based on a 100-point scale, you simply subtract 100 to compute the change. For example, an index of 120 means a 120 – 100 = 20-point improvement.

Here's the formula for a 100-point index:

```
(Current Performance
/ Comparison Performance)
* 100
```

I prefer a 100-point index to a percentage change formula because I don't like to compute a percentage of a percentage.

However, if you or your stakeholders are used to seeing the data in this format, you can easily convert the index formula to be a percent change formula by changing `* 100` in the preceding calculation to - `1`.

Now that we've isolated the current performance, the prior performance, and the comparison index, we're ready to make the dashboard element. The only way to combine different fields *and* different mark types in this dashboard element is to use a placeholder calculated field with the formula `MIN(0)` and place the calculated field on the Rows Shelf three times:

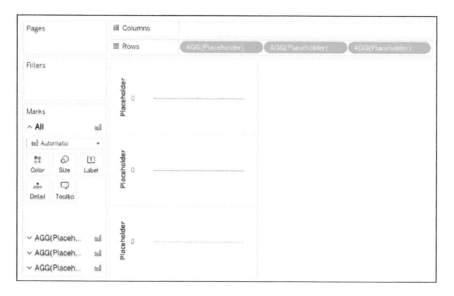

Now that there are three measures on the Rows Shelf, there are three different Marks Shelves that we can independently edit. This means that I can put Current Performance with a mark type of Text on the first row, Comparison Performance with a mark type of Text on the second row, and the Comparison Index with a mark type of Square on the third row:

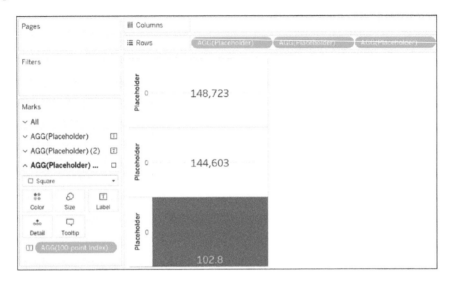

Lastly, I like to color the square based on the 100-point index score. I typically do one color for over 100 (outperformed), one color for a score of 90 to 100 (slightly underperformed), and one color for a score below 90 (underperformed). We can do this by using a calculated field such as the following:

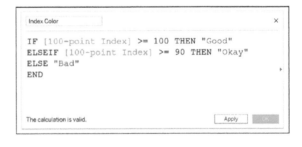

```
IF [100-point Index] >= 100 THEN "Good"
ELSEIF [100-point Index] >= 90 THEN "Okay"
ELSE "Bad"
END
```

This approach works OK when you're making this index callout for only a few metrics, but it can be a bit manual with all of the calculation creation. The coloring is much easier if you want to communicate only a positive or negative change like we demonstrate in Chapter 38; you would simply place the index score on the Color Marks Card and modify the colors so that they're on a diverging color palette, with a step size of 2, and a center of 0.

Related: *Practical Tableau*, Chapter 61, "How to Make a Stoplight 100-Point Index" (O'Reilly, 2018)

Finally, the only thing you lose with this placeholder hack is the title of each row. However, you can add that by modifying the text on the Text Marks Card. Here's how my final dashboard element looks after I tweaked the format, added sheet titles via the Text Marks Card, and colored the index score by my newly created Index Color calculated field:

An Alternative Current Versus Comparison Index Callout in Tableau

I like what we have so far because it displays the exact numbers for the current and comparison performance and it also introduces the preattentive attribute of color so that my stakeholders can see at a glance how they are doing across key performance indicators. But in the spirit of minimalism, lately I've been gravitating toward a slightly different look that collapses the three rows into one.

If you were following along, you already have the calculations and foundation required to create this second look. For this second approach, we show just the current performance and color the entire element based on the period versus comparison index.

One approach to creating the second view is to remove the first two rows from our first index callout, replace the Index Score on the label with Current Performance, and update the element title:

With this style, I like to add either the index point change or percent change following the current performance. As most are new to the 100-point index, this time I show a percent change. Here's the calculation:

```
([Current Performance] /
[Comparison Performance])
- 1
```

We then place this calculated field on the label by dragging it to the Label Marks Card. Let's also customize the label so that both numbers are on the same line by clicking the Label Marks Card and modifying the text:

In this case, the current versus comparison performance was up 3%, so the element turned blue for positive. If I were to change the comparison to May, when the period-over-period performance was negative, but within 10%, the element turns to yellow for Okay:

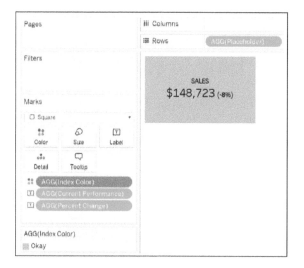

If I change the comparison to a month that July underperformed by more than 10%, such as March, the element turns a third color to indicate Bad:

Whether you prefer the first or second style, I recommend placing a series of these elements across the top or down the left side of your summary dashboard to provide a starting point for your explanatory analysis. After you've got it down, take the percent change formatting a step further by

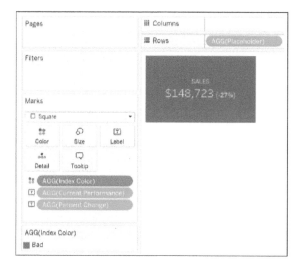

conditionally formatting the text, as we demonstrate in Chapter 48.

CHAPTER 40

How to Highlight a Dimension Member on a Minimalist Dot Plot

There is a chart type that I've been gravitating toward a lot but struggling to find a documented name for. The chart consists of *dots* (or the Circle mark type in Tableau) *plotted* on a shared axis. I've been calling this a dot plot for seemingly obvious reasons, but traditional dot plots are simply a bar chart with the mark type changed from Bar to Circle. My examples are slightly different in that they share one axis, or row of dots, and I usually hide the axis header. Being that this is a plot of dots and includes only the minimum amount of data possible, for now I'm calling this a *minimalist dot plot*.

This chart type has some big benefits. First, it's the closest I've seen to getting to a 100% data-to-ink ratio. It provides comparisons which help avoid the dreaded, "So what?" Also, by hiding the axis header, the user is forced to focus on the insight of comparing dimension members in relation to one another rather than the exact numbers. Of course, I provide the exact data points on demand via the tooltip when the user hovers over a circle.

The chart type is featured prominently in both my *Super Sample Superstore* (*https:// oreil.ly/Ea4dn*) and *MLS Standings Reinvented* (*https://oreil.ly/hSKDb*) dashboards. In the first, the user can choose which region they hypothetically manage; in the latter, the user can choose their favorite team. In both cases, the dot plots then highlight the selection throughout the dashboard so that the user can see where they stand in relation to the others.

This chapter shows you how to make a minimalist dot plot in Tableau and how to highlight a specific dimension member throughout multiple views.

How to Make a Minimalist Dot Plot in Tableau

This chart type is one of the easiest to create in Tableau and requires only a few simple steps. For the purposes of illustration, let's rebuild one of the dot plots from the *Super Sample Superstore* dashboard. Note the charts at the bottom of the view, which highlight based on the user selection:

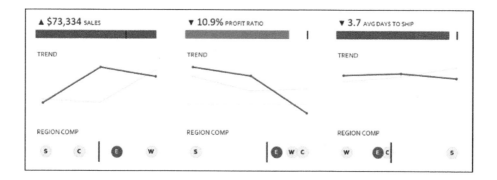

To re-create the Sales dot plot from the Sample – Superstore dataset, drag the Sales measure onto the Columns Shelf:

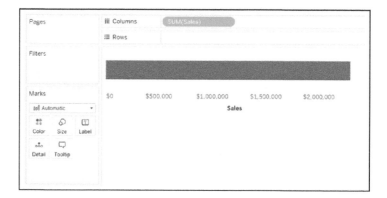

On the Marks Shelf, click the drop-down box and then change the mark type to Circle. Optionally, you can hide the axis by right-clicking it and deselecting Show Header:

The most critical step is to place the dimension that you want to compare on the Detail Marks Card. In this case, I am comparing regions, so I place the Region dimension on the Detail Marks Card. This changes the level of detail for the view and draws one circle for each region:

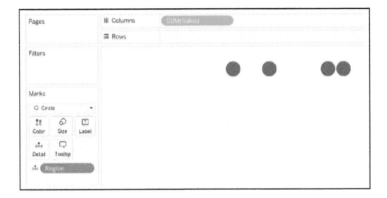

Optionally, you can also add a reference line by navigating to the Analytics pane and dragging Reference Line onto the view. Here's how my final view looks after I add and format a reference line for average, removing grid lines and axis rulers and changing the color of the circles:

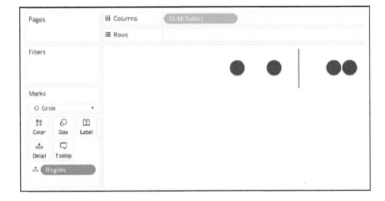

With this chart type, I care most about the marks' relations to one another, so I usually don't mind if the axis doesn't start at zero. I'm sure you've heard the argument for including zero, so this choice is up to you and your stakeholders.

How to Highlight a Dimension Member Throughout Dot Plots

The trick to highlighting a specific dimension member on one or more of these dot plots involves parameters. To begin, create a parameter with a data type of String. For the list of allowable values, click Add from Field and then, on the menu that opens, choose the dimension you are comparing:

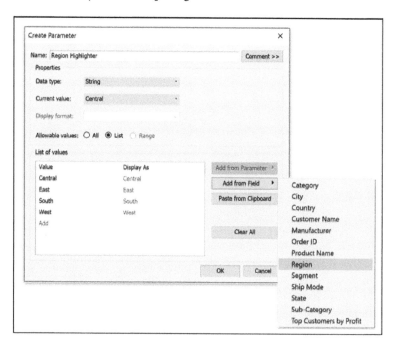

Next, create a calculated field with a Boolean formula. The calculation will be as follows:

```
[The dimension you are comparing]
= [The highlighter parameter]
```

Place this newly created calculated field on the Color Marks Card. If the dimension member matches the parameter selection, it will be colored a distinct color from the other dimension members:

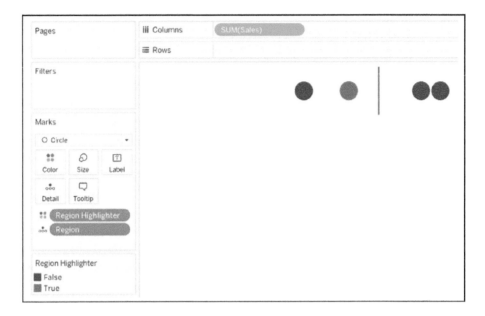

Don't forget that if you want your end users to be able to choose the parameter selection, you must show the parameter control by right-clicking the parameter and then selecting the Show Parameter Control checkbox. Now, you can use the highlighter calculated field to color marks across multiple dot plots, and choosing the dimension member to highlight on the parameter control will highlight the specific dimension member across all of them!

CHAPTER 41

3 Ways to Make Gorgeous Gauges: Tip 1

Whereas bullet graphs are the optimal type of *gauge* in Tableau because of their efficient use of space and their ability to show values past 100%, there are other engaging ways to display the progress toward a goal or prior period. The next three chapters show you how to make a variety of gauges in Tableau. For the purposes of this chapter series, I define *gauges* as chart types that show progress to a goal or comparison point.

If your primary objective is to communicate how much progress you've made toward hitting the 100% mark, and you don't mind not seeing performance past the goal, you can stop the scale at 100%. This lends itself to some interesting design possibilities including the oft-maligned donut chart. Donut charts are criticized for inefficiently using dashboard real estate, stopping at 100%, and making it difficult for users to accurately assess progress to goal.

This chapter shows you how to essentially flatten out a donut chart, which will solve two of these three deficiencies. Plus, I show you a hack that allows you to round bars and the background scales. This is not an out-the-box design in Tableau, but I think it adds a touch of engagement to gauges.

 Related: *Practical Tableau*, Chapter 26, "How to Make Bullet Graphs" (O'Reilly, 2018)

How to Make Rounded Gauges

By the end of this chapter, you will be able to make rounded gauges in Tableau that look like this:

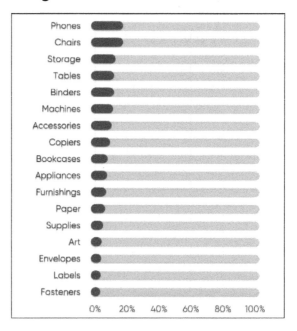

I'm using Percent of Sales by Sub-Category in the Sample – Superstore dataset, but you can use the following technique for any measures that you want to display on a 0 to 100% scale. Admittedly, Percent of Sales is not the best use case for this type of chart, but I want you to be able to follow along without having to provide a separate dataset. The reason Percent of Sales is not the best use case is because no dimension member will ever fill the gauge all the way to 100% and this would be just as effective as a basic bar chart. This type of gauge works best when you are wanting to track and communicate progress toward a goal or comparison up to, but not past, 100%.

To create rounded gauges in Tableau, you need three measures:

- The measure you are analyzing
- A calculated field that creates the bottom of the gauge
- A calculated field that creates the end of the gauge

The measure that we analyze in this example is a table calculation that computes the percent of sales across the 17 sub-categories in the Sample – Superstore dataset; this will always total to 100%.

The calculation for the start of my gauge should equal zero, so the formula is as follows:

The calculation for the end of the gauge should equal one; here's its formula:

Next, we need to get these three measures onto the view and break them down by, in this case, sub-category. To do so, let's use the generated field Measure Values on the Columns Shelf and the Sub-Category dimension on the Rows Shelf:

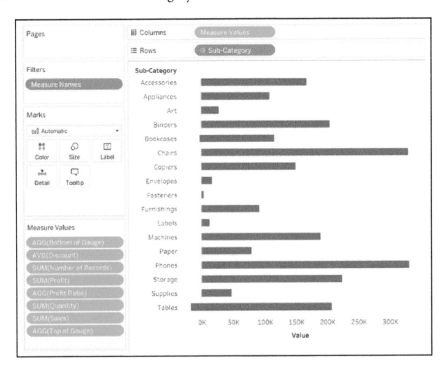

By default, this stacks every measure in my dataset in a bar chart, broken down by sub-category. For this visualization, we need only the Sales, Bottom of Gauge, and Top of Gauge measures, so let's drag the others away from the Measure Values Shelf to filter the measures being used:

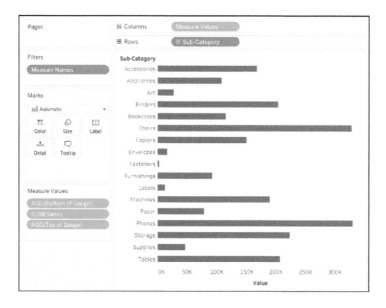

For this example, we need to add a table calculation to compute the percent of total sales for each sub-category. To do this, right-click the Sales measure that is currently on the Measure Values Shelf and then, on the menu that opens, hover over Quick Table Calculation, and then choose Percent of Total. Now all three measures are stacked on top of one another on a 100% scale:

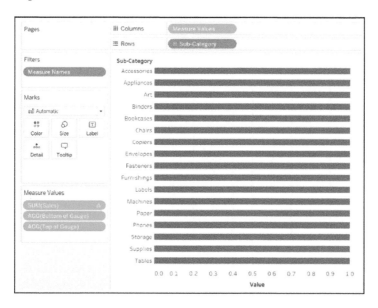

To see the three individual measures, drag the generated field, Measure Names, to the Color Marks Card. The bottom of the gauge and the measure you are analyzing should be colored the same, and the top of the gauge should be a different, more neutral color (this will eventually be the gauge background).

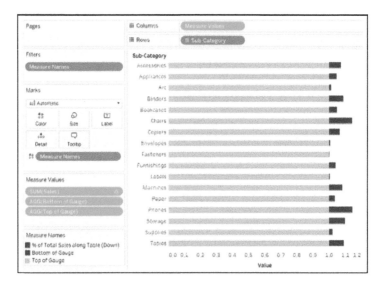

Now, you can see that the scale is going beyond 100%. To reset the gauge to again represent zero to 100%, you must turn *off* stacked marks. To do so, on the menu bar at the top, click Analysis > Stack Marks. After choosing Off, the chart looks like this:

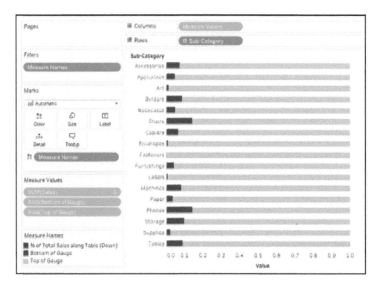

This is a decent chart at this point, but it has at least one drawback: the bars can be implied as stacked bars and confuse the user into thinking the gray shading means something other than blank space that is yet to be filled. Although it comes with its own limitations (most notably less precision), I believe rounded marks help imply that the chart is a gauge that is filling up.

To create the rounded design, duplicate the Measure Values pill on the Columns Shelf. I like to do this by holding down the Control key while I click the pill and drag it next to itself:

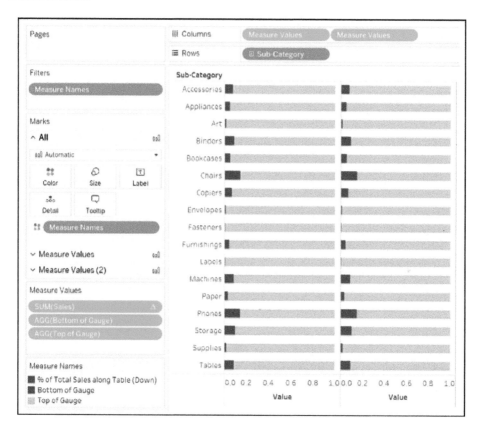

This puts the same gauges in two columns, but now that there are multiple measures on the Columns Shelf, they each get their own set of Marks Cards that we can independently edit. This means that I can navigate to the Marks Shelf for the second column and change the mark type to Circle:

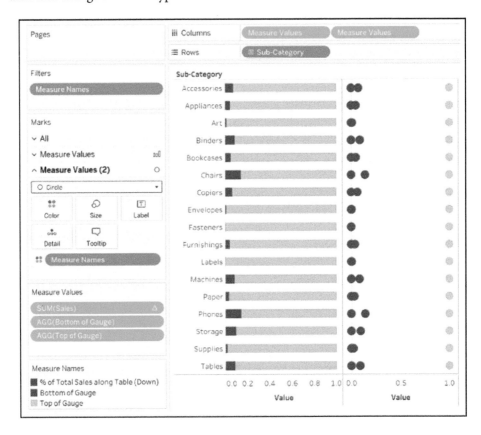

Next, convert these two columns into a dual-axis combination chart by right-clicking the second pill on the Columns Shelf and then choosing Dual Axis. Tableau will try to help you out and change the mark type of the first column to Circle. Make sure that you change the mark type of the first column back to Bar:

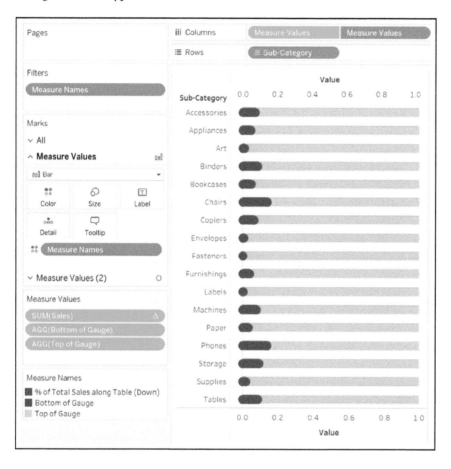

To finalize the view, use the Size Marks Card to get the circles to line up with the ends of the bars and ensure that the axes are synchronized. To do this, right-click either axis and then choose Synchronize Axis:

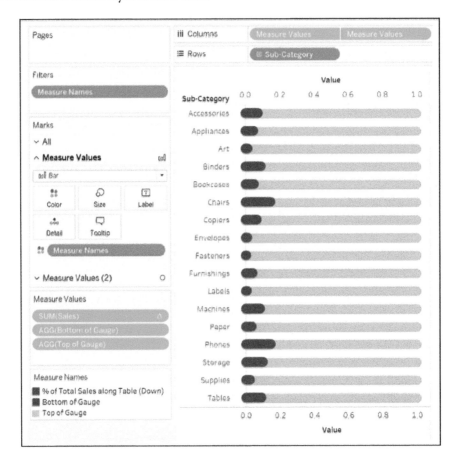

Here's how the final view looks after I hide one of the axes, change the number formatting to percentages, sort, and clean up the formatting:

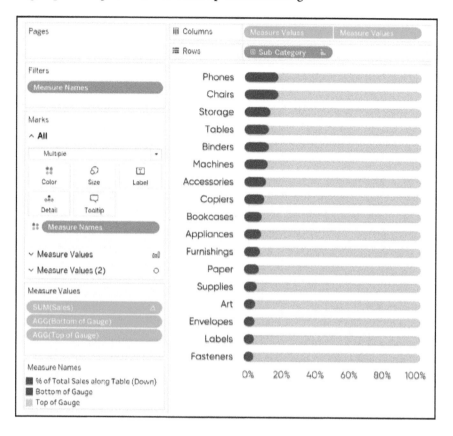

Not only do the rounded ends add a nice aesthetic to traditional bar charts or gauges, but in this case, they're serving the practical purpose of implying the gauges are filling up from left to right. Rounded gauges like these are an effective and engaging solution any time you are visualizing marks on a 0 to 100% scale.

CHAPTER 42

3 Ways to Make Gorgeous Gauges: Tip 2

How to Make Gauges with a Custom Background Image

This chapter shows you how to make a gauge with *any image*. I learned this trick from Tableau Zen Master Lindsey Poulter in her visualization, *Best States to Raise Children*. Downloading Tableau public visualizations and reverse engineering them is one of the best ways to pick up new techniques.

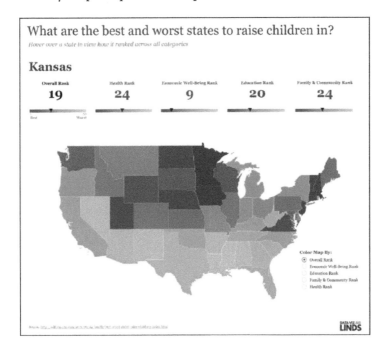

In the visualization, Lindsey draws an arrow over a custom sequential color palette to communicate performance across five different categories. In this tutorial, I show you how to make a gauge out of a custom background image for a company that is interested in viewing individual responses to a Net Promoter Score (NPS) survey. The following approach is useful any time you want to customize a gauge with an image of your choosing.

How to Customize Tableau Gauges Using Images

Here is an example of a gauge built with a custom background showing a response to an NPS survey with rounded shapes colored to indicate whether the respondent is a

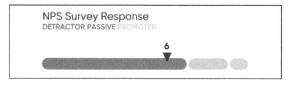

Detractor (responded between 0 and 6), Passive (responded 7 or 8), or a Promoter (responded 9 or 10).

Typically, NPS scores are calculated by taking the percentage of Detractors and subtracting that number by the percentage of Promoters, but for this illustration, let's pretend that we are wanting to look at individual responses. Again, you can apply the following techniques to any use case for which you want to create a custom gauge using an image.

Approach 1: How to Create a Custom Gauge in Tableau by Mapping a Custom Background

For the first technique, let's map the custom gauge pictured earlier using the traditional way of mapping a background image in Tableau, as you saw in Chapter 31. To lay the framework for our custom gauge, we need two measures:

- A placeholder field, which will create the y-axis. For this arbitrary placeholder measure, I always create a calculated field with the following formula:

- The metric we want to show up on the custom gauge. This will be plotted on the x-axis, or Columns Shelf.

I can show you both measures using the Sample – Superstore dataset because the first is a calculated field that I can create in any dataset, and I parameterize the second to represent different NPS responses. To create the second measure, I set up a parameter with a data type of Integer, allowable values of 0 to 10, and then created a calculated field that contained nothing but the parameter value. I'm doing this purely to show you an example using the sample data, but the second measure will be in your dataset, and you don't need to worry about these steps.

Now, let's start the view by placing the Placeholder measure onto the Rows Shelf, and the measure you are analyzing onto the Columns Shelf. By default, this creates a scatter plot with one data point on the view:

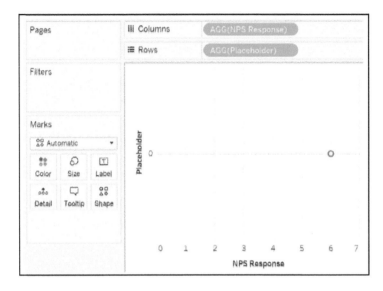

To map a custom background image in Tableau, on the menu bar at the top of the window, click Map > Background Image > [Your Data Source]:

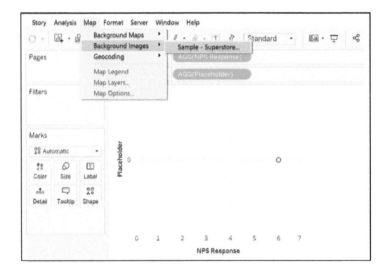

Click Add Image to open the Add Background Image dialog box in which you can find the custom image on your computer and map the x and y axes. The measure selected and values for the X Field determine how far the custom image spans *horizontally*; the measure selected and values for the Y Field determine how far the custom image spans *vertically*.

For this example, we want the image to go left to right for values between 0 and 10, so let's set those values for the X Field. The values for Y are more arbitrary, but we can use them to determine how much vertical space the image represents. Let's set it to –1 to 0 for now. We do this because the NPS measure will always be plotted vertically at the value of 0, so in this case, the mark will be displayed at the *top* of the gauge. For best results, click the Options tab and then select the Always Show Entire Image checkbox. Here are the final settings:

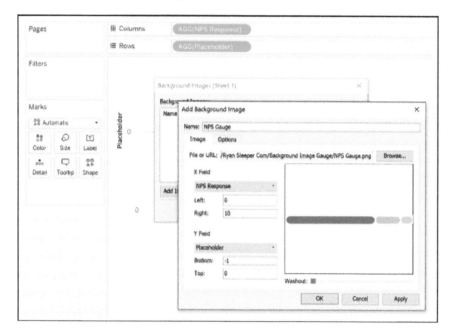

After you click OK twice to close the settings, you will see the custom image starting to come into focus:

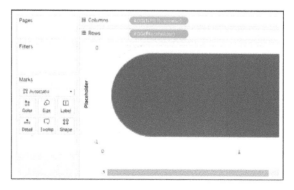

Due to this image's resolution, selecting the option to always show the entire image causes part of the image to be pushed off the screen. To alleviate this, on the toolbar at the top of the Authoring interface, click the drop-down box and change the fit of the view to Entire View. We will resize this gauge after it is added to a dashboard anyway.

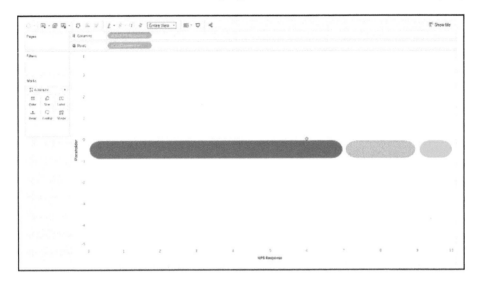

All that's left to do from this point is formatting! Here's how the view looks after I hide all lines, hide both headers, choose a triangle shape for the mark, and add the NPS score to the Label Marks Card:

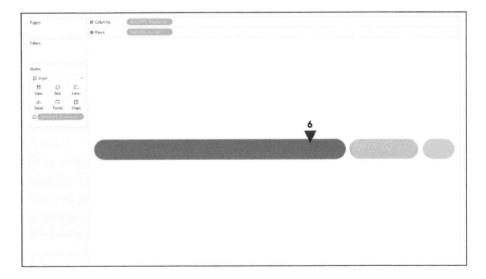

Here's how the element looks after I add it to a dashboard as a floating object so that I can control its dimensions:

Because I parameterized the NPS responses, I can show you how the view looks if I change the current value of the NPS parameter to 8:

Regardless of what response you choose to populate the value on the x-axis, the slider will move to the appropriate place on top of the custom gauge in the background! The colors help the viewer instantly identify what segment the respondent falls in and the magnitude of their brand affinity. If you lined several of these gauges up on a dashboard, this could also be used to compare respondents to one another to see at a glance who are the biggest Promoters or Detractors across different areas of the business.

Approach 2: How to Use Transparent Sheets to Create Custom Gauges in Tableau

As mentioned in Chapter 5, you can make the backgrounds of sheets transparent. This unlocks even more design flexibility in Tableau, including the option to float a sheet over an image of a custom gauge. If you're using Tableau version 2018.3 or later, you can take advantage of this technique, which requires fewer steps than the first approach. The trade-off is that it's not quite as precise because the background image is not directly tied to the values on the view; you are simply lining up a sheet on top of an image.

To somewhat alleviate this drawback, you want to ensure that you have fixed the range of the x-axis from 0 to 10. This way if you make the worksheet and custom background image the same number of pixels wide, they will be on the same scale.

To make a dashboard sheet transparent, right-click anywhere on the view and then, on the menu that opens, choose Format. For the shading, make sure the worksheet shading is set to None:

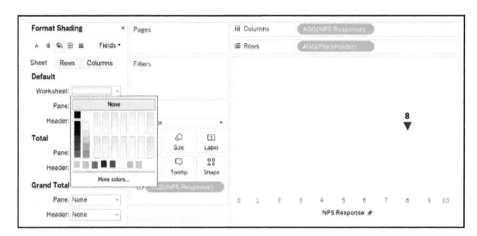

To create a custom gauge using transparent sheets, create a new dashboard and add an Image object that contains your custom gauge image. This works best when the objects are floating because you can precisely size them to ensure that everything lines up on the same scale:

Next, float the transparent sheet over the Image object. Note that for this example I've rehidden all the lines:

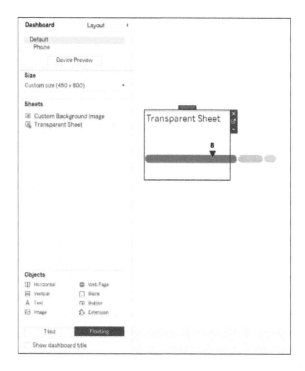

As you can see, they're not yet lined up because the transparent sheet and Image object have different widths and x coordinates. You can change the settings of a floating object by navigating to the Layout pane. After you ensure that both the transparent sheet with the NPS response and the custom gauge image are 359 pixels wide and start at the x-position of 33, the gauge lines up again:

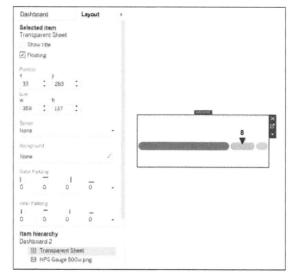

Whether you choose approach one for its precision and ability to tie data points directly to an image, or approach two for its ease of implementation, integrating images into gauges is a great way to improve your design and engage users!

CHAPTER 43

3 Ways to Make Gorgeous Gauges: Tip 3

How to Make a Stock Ticker Gauge

The next gauge I show you how to build in Tableau was inspired by the stock tickers seen on major news networks and stock portfolio apps. The visualization typically features the positive or negative change of a stock or index, preceded by an up or down colored triangle, and are occasionally enclosed by a colored rectangle to reinforce the change.

Although stock tickers aren't the best choice for visualizing the *magnitude* of a change, I like them because they clearly indicate a positive or negative performance, they have a minimalist design, and they are encoded by both color and shape; an approach to double-encoding that is colorblind friendly. If you're looking to try a slightly different approach, the stock ticker gauge is a good alternative to the performance indicator titles with comparison scores that are outlined in Chapter 39. This chapter shows you how to use Gantt charts in a unique way and employ the up (▲) and down (▼) alt code characters to create a stock ticker in Tableau.

How to Use Gantt Marks and Alt Code Characters to Create Gauges

By the end of this chapter, you will be able to make a stock ticker gauge with an up or down triangle and rectangular indicator that changes color based on performance:

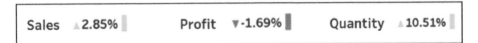

This gauge requires four elements:

- A calculated field that computes the comparison between the current performance and your comparison point
- A calculation that creates the up triangle
- A calculation that creates the down triangle
- A Gantt mark (optional)

The Gantt mark is mostly aesthetic, but when used in a series displaying multiple KPIs, it has the practical purposes of enclosing the gauge and providing a visual separation. This is similar to how column dividers separate text in crosstabs or how pipe symbols (|) provide a visually appealing separator. But in Tableau, using a bit of trickery, we can give the mark more value by dynamically coloring it based on performance. You can also use the Gantt mark by itself instead of triangles, but you lose the encoding by shape which helps colorblind audiences.

 This element is optional. If you prefer to display only the text and triangles, skip down to the next element of the gauge and place the calculations on the Text Marks Card.

To lay the foundation for our stock ticker, let's create what I refer to as a *placeholder* field. I often create a calculated field for my placeholder, but this time, I'll show you how to make a placeholder field in the flow (see Chapter 61).

On a new sheet, double-click the Columns Shelf, type **MIN(0)**, and then click Enter. Even though the number 0 technically does not need an aggregation, it will be important to have an aggregation of minimum later if we change this value to properly align our Gantt mark:

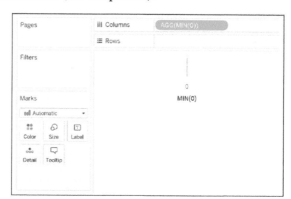

Next, using the drop-down box on the Marks Shelf, change the mark type from Automatic (Bar) to Gantt Bar:

Gantt bars must be sized by something, and whatever value we use to size our Gantt mark will end up being the size of our colored rectangle indicator. This is completely arbitrary, but let's size our mark by the value of 1. This is another calculation that you can build in the flow by double-clicking any blank space on the Marks Shelf, typing **MIN(1)** and clicking Enter.

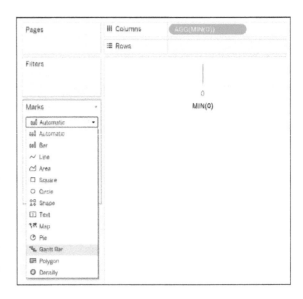

By default, this new calculation will be placed on the Detail Marks Card. If you drag it from its default location to the Size Marks Card, the Gantt bar (i.e., our indicator) will be sized by 1:

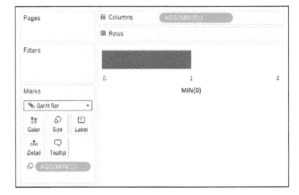

As you can see, the Gantt mark is spanning from 0 to 1. From here, you can edit the value we used to create the placeholder on the Columns Shelf and or the axis range (by right-clicking the axis and then choosing Edit Axis) to resize the indicator. Here's how mine looks after changing the placeholder value to 9 and my axis range from –10 to 10:

You will also want to remove all the lines. Right-click anywhere on the view and then, on the menu that opens, choose Format, and then click the Lines tab (the three stacked lines) and set Grid Lines, Zero Lines, and Axis Rulers to None:

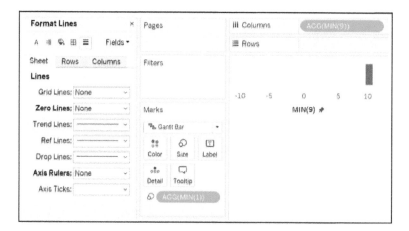

We've laid the foundation for our stock ticker gauge without even creating a calculated field so far. The rest of the gauge is completed by adding the change value and triangles as labels to the Gantt chart being used as our indicator. I personally like the effect the mark creates, but I admit that if you're using triangles in addition to the Gantt mark, this is not required and sometimes the "juice is not worth the squeeze."

So, I don't blame you if you skip that foundational step, but you will still need the following calculations to create a stock ticker gauge without the Gantt mark. If you prefer to make something similar, but exclusively using triangles, simply start a new worksheet and place the following calculations on the Text Marks Card instead of the Label Marks Card.

First, the change calculation. For ease of illustration, I have parameterized a calculated field, but you will likely need to create a calculation that computes either a period-over-period change or comparison to goal. The calculation for percent change is ([Current Performance] / [Comparison Performance]) - 1.

We then need a calculated field that creates an up triangle and one that creates a down triangle. These need to be separate for two reasons:

- They'll allow us to independently format the triangles and the change value on the label (so the triangles can be a smaller font size)
- We can color the up and down triangles different colors (for additional context, see "How to Conditionally Format Text" on page 285)

Here's the calculation for the up triangle:

IIF([Change]>0,'▲',NULL)

This is saying that if the change is greater than zero, display the up triangle; otherwise the value should be null.

And here's the calculation for the down triangle:

IIF([Change]<0,'▼',NULL)

This is saying that if the change is less than zero, display the down triangle; otherwise the value should be null.

We're now ready to place all of our fields onto the Label Marks Card of our Gantt chart (or Text Marks Card if you skipped ahead). After placing the calculated fields on the appropriate Marks Card, format the label as appropriate. I prefer to make my triangles smaller than the change value and color the up triangle blue and the down triangle red:

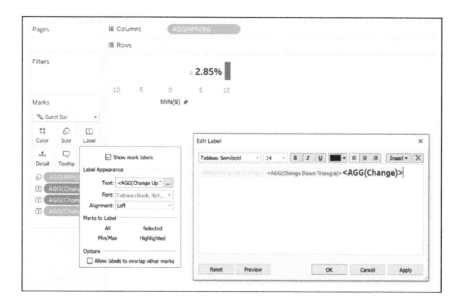

The way that our triangle calculated fields are set up, only one triangle will display at a time. When the change is positive, the up triangle will display and the down triangle will be null; when the change is negative, the down triangle will display and the up triangle will be null.

Next, place the Change measure that you are using onto the Color Marks Card; this colors the rectangular indicator. Modify the diverging color legend (by double-clicking it) so that the stepped colors are 2 and the center is 0; this displays one color for positive values and a separate color for negative values. I recommend that you make these colors match the colors used for the triangles:

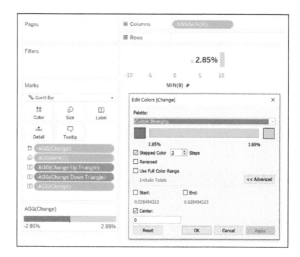

To finalize the view, format the number of decimal places to your liking, use the Size Marks Card to change the height of the rectangle and/or hide the axis by right-clicking it and then deselecting Show Header:

There are several ways to get the measure name to which the stock ticker gauge is related onto the view (adding a sheet title, adding a text object to a dashboard, using the Measure Names field and editing the alias, just to name a few).

But here's one more sneaky idea. You can add the measure name by adding a calculation in the flow just like we did earlier with the placeholder field. Here's how my final view looks after I double-click the Rows Shelf, type Sales, and then click Enter:

This is an engaging explanatory visualization communicating at a glance which KPIs in the business need further attention. Not only does it have a nice, minimalist design, but the double-encoding by shape and color help aid understanding.

CHAPTER 44

How to Make Ranged Dot Plots in Tableau

Ranged dot plots display not only a circle mark, or dot, representing the current performance for a specific dimension member, but a full range of how that dot has moved over time. You can add even more context by coloring the dot and/or changing the mark type of the circle to up or down triangles based on period-over-period performance.

Both of these additions to a traditional dot plot help provide comparisons that paint a clearer picture of whether changes in performance are notable. This chapter shows you how to add a performance range for each dimension member represented in a dot plot.

How to Make a Dot Plot with a Performance Range for Each Dimension Member

By the end of this chapter, you'll be able to make a dot plot showing the period-over-period performance across dimension members with a performance range for each dimension member in the background:

Let's knock out the traditional dot plot portion of this chart first. You can create this chart with any measures and dimensions, but, here, we look at current month's sales by region in the Sample – Superstore dataset.

For ease of illustration, I have parameterized the current month selection so that the user can choose any month between January and December as the *current* month. Also, the sample data has four years of data in it, but we will aggregate all four years into one (i.e., pretending there's only one year of data).

The first thing we need to do is isolate the performance for the current month. Here's the formula for the fields:

```
SUM(
IF [Parameters]
.[Current Month] =
MONTH([Order Date])
THEN [Sales] END)
```

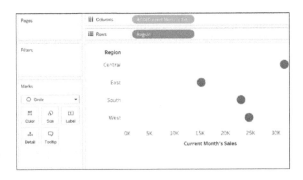

To analyze this measure by the Region dimension as a dot plot, place the newly created calculated field on the Columns Shelf, the Region dimension on the Rows Shelf, and change the mark type to Circle:

By default, the current value of a parameter is the first allowable value. Because the first allowable value in this case is 1, so far we are looking at January's sales by region.

If we show the parameter control and change the value of the parameter to 8, the value of 8 will replace the value of 1 in my Current Month's Sales calculated field, and we will be looking at August's sales per region:

The dot plot is already providing one comparison: Region. The next step is optional, but I typically like to add at least one more comparison to a dot plot by coloring each circle based on the period-over-period performance of each dimension member. You could also change the mark type to Shape and map up triangles for positive changes and down triangles for negative changes.

In either case, we need to make a calculated field that determines the month-over-month change. In this case, the formula is as follows:

```
SUM(
IF [Parameters]
.[Current Month] =
MONTH([Order Date])
THEN [Sales] END)
-
SUM(
IF [Parameters]
.[Current Month]-1 =
MONTH([Order Date])
THEN [Sales] END)
```

Now, if we place this calculated field on the Color Marks Card, change the colors to a diverging palette, and map one color for positives and one color for negatives, we can quickly compare performance between dimension members and between months:

Now for the new stuff.

The trick to creating a ranged dot plot is to combine a traditional dot plot with a Gantt chart representing the performance range for each dot. The Gantt mark for each dimension member will start at the beginning of the range, so we need to isolate the value for the worst performing month per row. Here's the formula:

```
MIN({FIXED [Region],MONTH([Order Date]): SUM([Sales])})
```

We're using a level of detail expression, but don't let this formula intimidate you! Just think logically through what we're trying to isolate: the lowest value, aggregated at the monthly level, for each region. When that's the case, my brain first thinks of MIN([Sales]). Well, in our view, the level of detail is Region, so the Sales measure with this aggregation will take the lowest sales value per region.

This isn't quite what we want because several orders make up each month's sales, and in this case, Tableau would display the lowest *order* value for each region (but not for each month). What the FIXED level of detail expression does is aggregates those multiple rows with all our different orders at whatever level you address within the calculated field. I have rolled up all the SUM([Sales]) values at the Region and Monthly levels.

Level of detail calculations are unique in Tableau in that you can create an aggregate of an aggregate. So, after summing up sales per region and month, I've wrapped that value in an aggregation of MIN to derive the lowest monthly sales value per region.

 Related: *Practical Tableau*, Chapter 16, "An Introduction to Level of Detail Expressions" (O'Reilly, 2018)

To get this value on the view as a Gantt mark, simply place the calculated measure on the Columns Shelf and change the mark type to Gantt Bar. I've also removed the field on the Color Marks Card for the right side:

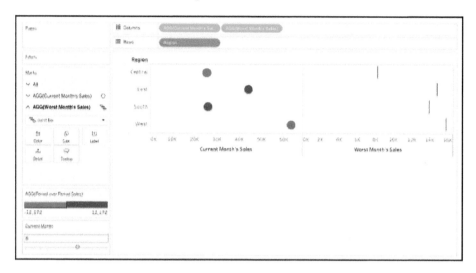

To size the Gantt marks so that they represent the full performance range from the worst month's sales to the best month's sales, we need to size them by that range. To compute that range, we need to know the value for the best performing month for each row.

We already created the calculated field isolating the worst performance, so we can simply duplicate that calculation and change the MIN aggregation at the beginning of the formula to MAX:

```
MAX({FIXED [Region],
MONTH([Order Date]):
SUM([Sales])})
```

The last calculated field we need takes the difference between the highest sales value and lowest sales value. For our measures, the formula is as follows:

```
[Best Month's Sales]
- [Worst Month's Sales]
```

You could consolidate these two calculated fields into one, but if you want to use Best Month's Sales somewhere else in the future, there is value in keeping them separate.

Place the calculated field computing the difference between the highest month's sales and lowest month's sales on the Size Marks Card for the Gantt chart:

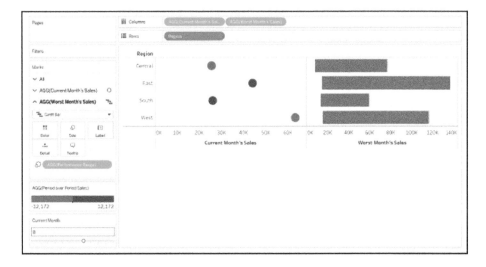

There are a few short steps to finalize the ranged dot plot:

1. Convert these two separate charts into a dual-axis combination chart by clicking the second pill on the Columns Shelf and choosing Dual Axis.

2. Synchronize the axes by right-clicking either one and choosing Synchronize Axis.

3. Move the circles in front of the Gantt bars by right-clicking the axis for the current performance and then choosing "Move marks to front."

4. Reduce the height of the Gantt marks by navigating to the Marks Shelf for the Gantt chart, clicking the Size Marks Card, and dragging the slider to the left.

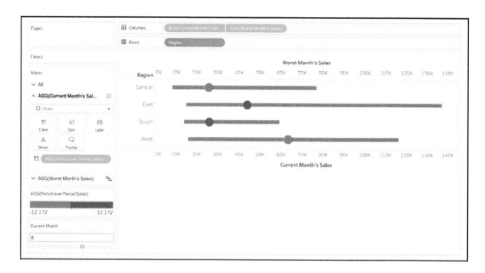

Here's how the final ranged dot plot looks after formatting:

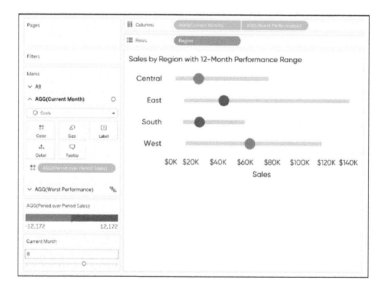

In this final view, the circle's position on the x-axis represents its current month sales, the color represents month-over-month sales, and the gray bar represents the range of values the dimension member has achieved during the year.

If I were to change the current month, the circle's position and color would change, but the gray bar is always frozen in the same place (unless the data is refreshed and the range changes). The gray bar in the background is adding value by providing additional historical context for each dimension member on the view.

CHAPTER 45

Introducing Leapfrog Charts in Tableau

Leapfrog charts are a variation on the minimalist dot plots described in Chapter 40. You can use them when the primary objective is to show the relative performance of a specific dimension member to a comparison point or points—or what it would take for that dimension member to *leapfrog* over another, if you will. The dot plot is combined with a Gantt chart to illustrate the difference between a selected dimension member and a specific target or the average, median, minimum, or maximum for a measure across the business.

This chart was developed in conjunction with Playfair Data information designer Jason Penrod. Keep reading to see how the chart looks and learn how to make it!

How to Combine Dot Plots, Gantt Marks, and Table Calculations to Create Leapfrog Charts

By the end of this chapter, you will be able to re-create this chart, which looks at Sales by Month by Region in the Sample – Superstore dataset and highlights the East region compared to the average for each month:

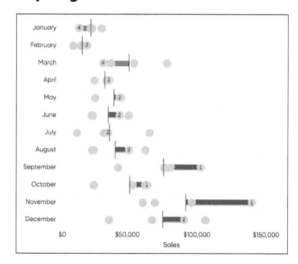

The first step to creating leapfrog charts in Tableau is to create the dot plot portion of the visualization. You can use any measure to create the columns, any dimension to create the rows, and any dimension to create the circles, but to illustrate, I use Sales, Month of Order Date, and Region from the Sample – Superstore dataset, respectively.

To create the dot plot:

1. Place the measure on the Columns Shelf.

2. Place the dimension that creates the rows on the Rows Shelf.

3. Change the mark type from Automatic to Circle.

4. Place the dimension that you want representing the circles on the Detail Marks Card.

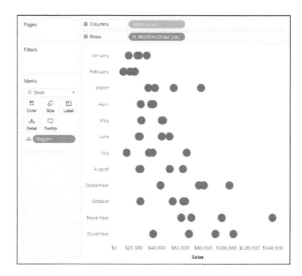

Now we need to create the Gantt chart that creates the bars in the background connecting the dimension member on which you are focusing to the comparison point. In this example, I'm focusing on the East region and comparing its performance to the overall sales average per month. To get the Gantt mark in the appropriate place on the axis, you must isolate the performance for the dimension member being focused on with a calculated field.

One way to do this is to hardcode the dimension member of interest. Here's the formula:

```
SUM(IF Region = 'East'
THEN [Sales] END)
```

However, I suggest parameterizing the region selection so that you can change the focus of the visualization on the fly.

Here's a Tableau tip within a Tableau tip: to create a parameter with a list of allowable values that matches the dimension members of a specific dimension, in the Dimensions area of the Data pane, right-click the dimension (i.e., Region) and hover over Create on the menu that opens to choose Parameter. This creates a parameter with the proper data type and populates your allowable values:

Note the default current value is the first allowable value (i.e., Central), so we are now focusing on the Central region until the value is changed in the parameter.

Let's replace the hardcoded East in the Focus Sales calculated field with this new parameter:

```
SUM(
IF Region =
[Region Parameter]
THEN [Sales]
END)
```

To use this calculated field to create the Gantt mark, place it on the Columns Shelf in front of the measure that is already there and then change its mark type from Automatic to Gantt Bar. By placing the focus measure in front of the measure that was already on the Columns Shelf, it just ensures that the Gantt marks are behind the circles in the final result:

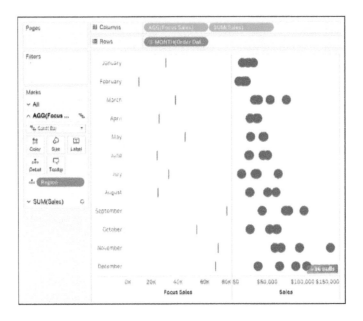

Convert this two-column chart into a dual-axis combination chart. Ensure that the axes are synchronized by right-clicking either axis and choosing Synchronize Axis:

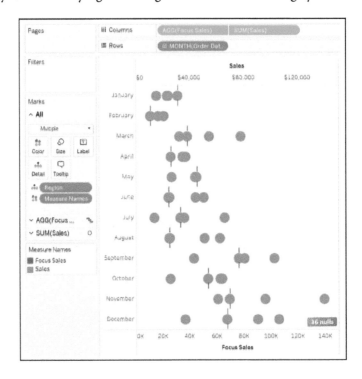

Because the comparison for this particular use case is average, let's go ahead and add a reference line for average per month to the Sales by Region axis. This will be helpful later when we quality check that the size and color of our Gantt marks are working as expected:

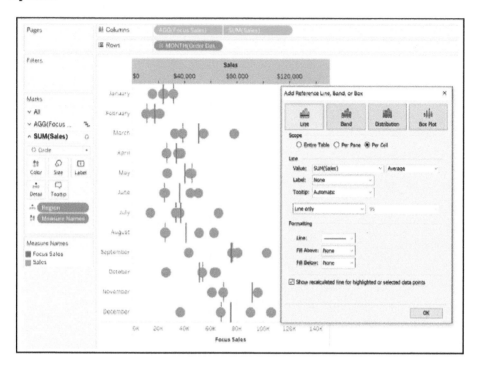

At this point, we have a circle for every region, a blue Gantt mark for the selected region, and a reference line showing us the average sales across all regions per month. Our focus region is beginning to get lost, so let's create a calculation to color the focus region something different from the other regions.

The most elegant way to write this calculation is as follows:

```
[Region] = [Region Parameter]
```

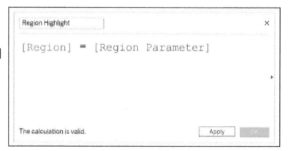

To create the highlight effect, replace the Measure Names field on the Color Marks Card of the dot plot with this newly created calculated field:

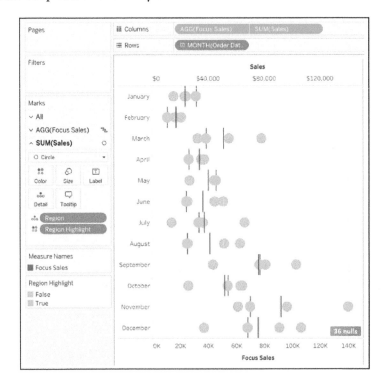

The Gantt marks need to be sized by the difference between the performance of your focus dimension member and the comparison, so we need another calculated field. This is the calculated field that really makes this chart "go." You can hardcode a target, isolate the performance of a benchmark, or, as we're doing in this use case, use window calculations to compare the dimension member to the average, median, minimum, or maximum for the measure being used.

Window table calculations are special in that they work across two axes, which we will need when using a dual-axis dot plot/Gantt chart. To compare the performance of the selected region to the average sales across sub-categories, the formula would be:

```
WINDOW_AVG(SUM([Sales])) - [Focus Sales]
```

Place this calculated measure on the Size Marks Card for the axis drawing the Gantt chart:

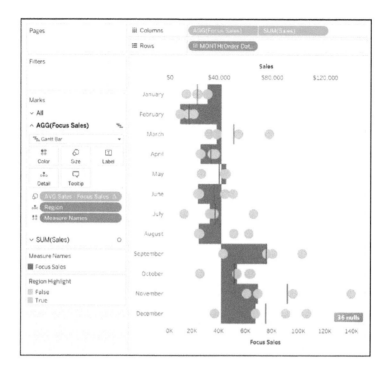

By default, this table calculation is moving from the top of the table down, so even though we are seeing the connecting lines come together, we need to change the addressing so that it includes the Region dimension and restarts the comparison every month.

This can be a little tricky to get right, but it is helped significantly by using calculation assistance. You can change the addressing of a table calculation and see calculation assistance by right-clicking the measure with the table calculation (currently on the Size Marks Card of the Gantt chart) and then choosing Edit Table Calculation. Note that you can also reorder the dimensions in the table calculation by just dragging and dropping them in front of each other within this dialog box.

In this case, when the table calculation is updated properly, there should be a Gantt mark connecting the center of the focus dimension member and the average for each month:

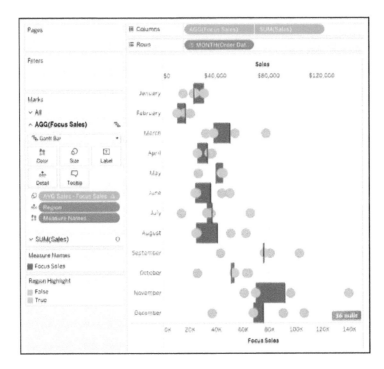

The Gantt mark should also be colored by this same calculation. The easiest way to do so is to navigate to the Marks Shelf for the Gantt Bar, hold the Control key, and click the measure containing the table calculation we set up during the last step while you drag it to the Color Marks Card.

Holding the Control key while clicking a pill creates a duplicate of that pill, including any table calculations that might be applied. This is handy because we don't need to edit the table calculation again to get the desired result:

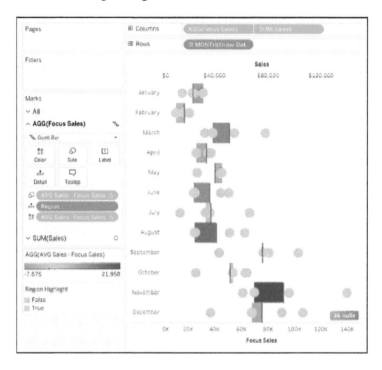

The Gantt bar colors don't look quite right at this point because when my focus dimension member is past the comparison point, we should see a color that represents positive performance; otherwise we should see a color that represents negative performance. To flip the color scale and reduce the number of colors included in the spectrum, simply double-click the color legend and change the settings:

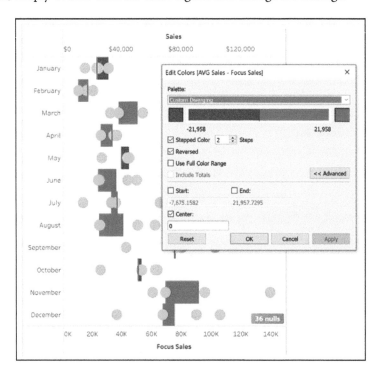

From here, it's all about the formatting.

At a minimum, I recommend making the size of the Gantt marks skinnier and the circles slightly larger by clicking their respective Size Marks Cards. I've also hidden one of the axes, hidden the indicators, turned off grid lines for Columns, changed the color of the reference lines, renamed the bottom axis, and put the fonts in brand. Finally, I changed the current value of the parameter to East to change the focus of the chart:

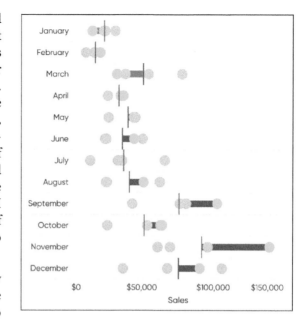

At this point, we can already determine at a glance how the East region did compared to the comparison point of average sales per month. We can also see the contextual indicators of how it did versus the other regions in sales per month. One interesting insight is that the East region started the first three months of the year behind average, and not only passed the average in April, but widened its gap with the comparison as the year progressed.

To add even more insight to this chart type, you can add a label to the circles that indicates rank. To do so, create a calculated field that computes the rank of SUM of Sales. You can also do this calculation in the flow directly on the Marks Shelf. In either case, the formula is RANK(SUM([Sales])) and should be added to the Label Marks Card:

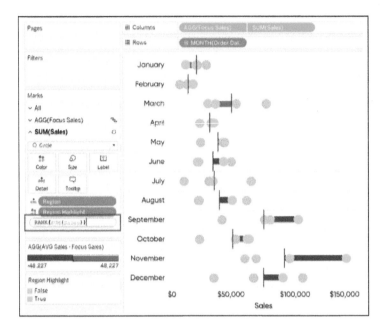

Rank is another table calculation, so we need to ensure that all the appropriate dimensions are being included, just as we did earlier with the first table calculation, for it to correctly compute the result. In this case, we want to see the sales rank computed across all our dimensions and restarting every month:

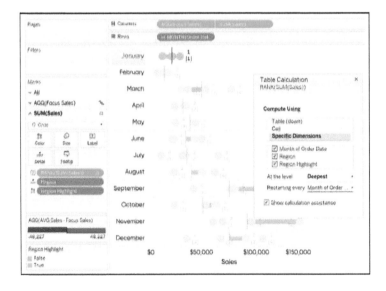

Let's center the labels on each circle by changing the Label Marks Card settings:

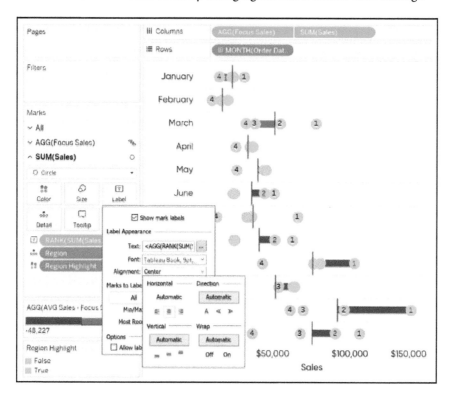

You have the option to show only the label for the focus dimension member, as you saw in this chapter's first image, but you will have more flexibility if you show all the labels. For example, because we parameterized the region selection, I can instantly change the focus of this visualization to the South region and see a completely different story unfold:

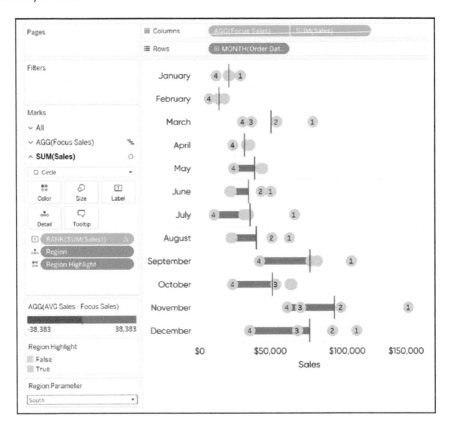

You can also change, or even parameterize, the dimension that is drawing the rows. However, you would need to update the table calculations to include that dimensional breakdown. Here's how the view looks after replacing the Month of Order Date dimension with the Sub-Category dimension on the Rows Shelf:

Leapfrog charts are an engaging alternative to dot plots alone and are packed with context. In addition to showing each dimension member's relative performance on the x-axis, we can now see how a focus dimension member compared to the average, the magnitude of that difference, and its rank compared to the other dimension members on the view.

CHAPTER 46

How to Let Users Choose Between Chart Types

When I was a kid, I enjoyed reading *Choose Your Own Adventure* books. In the books, after every few pages, the reader is presented with choices on how they want to proceed. Each choice will point you to a different page number where a story unfolds based on your selection. I assume that most kids reading the books were like me and would cheat by going back and forth to experience all the different outcomes.

Data visualization can be similar in that looking at the same data in different ways often leads to new insights (or storylines, if you will). Further, some end users will have their own preferences for how they want to look at data. Has anyone ever had to convert a data visualization to a crosstab view?

We've covered several charts during this section, but now it's time to let our users choose their own adventure! This chapter shows you how to empower your end users to choose which chart type to display, allowing them to toggle between multiple visualizations with the click of a button or selection from a drop-down menu.

How to Change Charts Using a Parameter Control

The trick to letting your end users choose the chart type they want displayed involves a layout container and—as is often the case with my Tableau tutorials—parameters. For the purposes of this tutorial, let's once again re-create a tactic from my visualization, *A Tale of 50 Cities* (*https://oreil.ly/ljqfu*).

In the visualization, the end user has the option to toggle between a highlight table and small multiple maps seamlessly in a single view (note the filter in the upper-left corner of the chart):

First, make your individual views as you normally would. My example has a highlight table and a map view, but you can create more than two options.

Next, create a string parameter with a list of choices that correspond to the individual views:

After creating the string parameter, create a calculated field that contains nothing but the newly created parameter. This formula acts as a binary; so, the calculated field will either match the parameter selection or it won't:

```
[Select View]
```

Now, let's use this calculated field as a filter. To get this to work properly, you will need to change the value of the parameter to match the view that you want displayed before adding the filter. As you can see in the preceding image, the current value of the parameter is Maps, so we can go ahead and add the Select View Filter calculated field to the Filters Shelf on the Map view and select the checkbox for Maps:

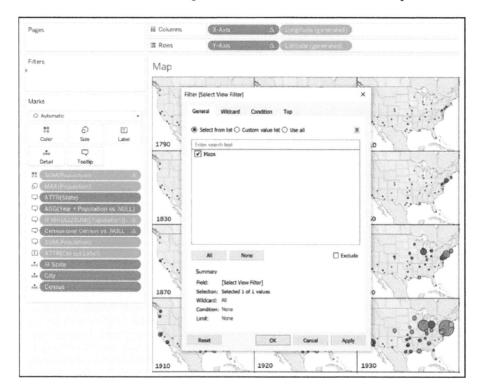

For the highlight table view, we will need to change the value of the parameter before adding the Select View Filter calculated field as a filter. To do this, right-click the parameter and then, on the menu that opens, choose Edit, and then change the current value:

After changing the current value, you'll notice that the first view disappears. So far so good. Remember the filter is a binary, so if the parameter selection doesn't match the calculated field, the view will be filtered out.

Now that we've changed the current value to the second chart type, we can add the Select View Filter calculated field to the Filters Shelf on the second view (which is the highlight table in this case):

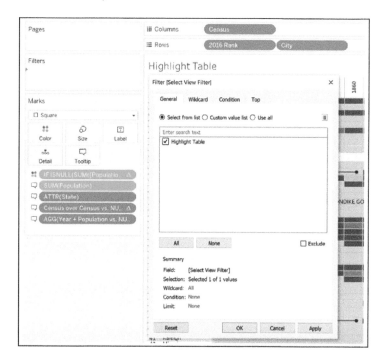

Now we have two individual views: one has a filter that will show only the view if Maps is the current value in the parameter, and the second has a filter that will show only the view if Highlight Table is the current value in the parameter. We're now ready to set up the view.

This is where the magic happens. Set up a dashboard that contains a vertical layout container (which is where we'll place the individual sheets):

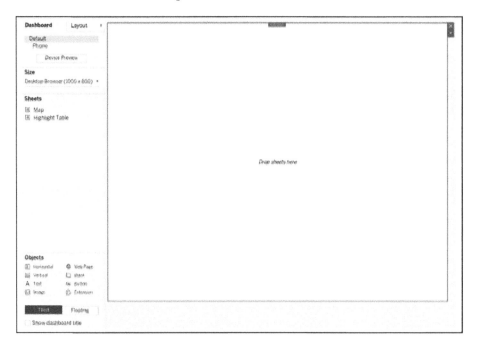

Now place both sheets into the layout container, one on top of the other. For best results, hide the titles by right-clicking the sheet titles and then choosing Hide Title. Notice that after adding the individual sheets to the layout container, only one view is shown. That's because we are filtering the views, showing only the view that matches the parameter selection, and you can make only one parameter selection at a time:

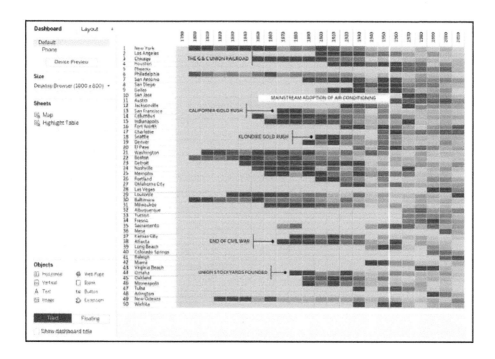

To make the parameter selection (i.e., choice of which chart to show) available to you and your end users, on the menu bar at the top of the window, click Analysis > Parameters and then choose the parameter:

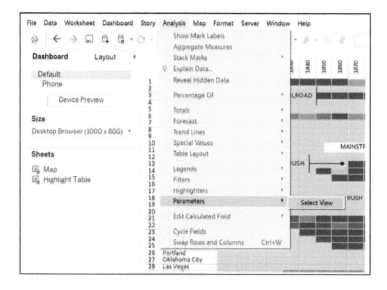

Now when you toggle between the parameter choices, only the appropriate chart will be shown. Here's the view when Highlight Table is selected:

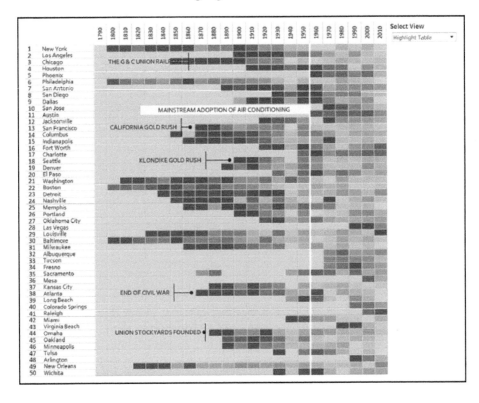

Here is the same view when Maps is selected:

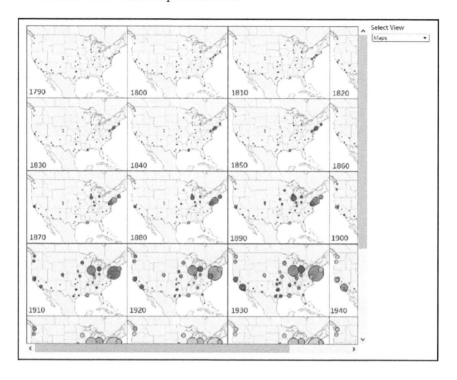

I kept the default fits and dashboard dimensions for this tutorial, but from here you can format the view, change the dimensions, float the parameter control, and so on to create your desired look and feel. As long as you keep the individual sheets in the layout container, the functionality will work. The only side effect is a small amount of white space added at the bottom of the layout container. With only two chart types, it's hardly noticeable, but it's something to consider if you want to try this out with additional views.

How to Change Chart Selections with Layout Containers

Now that we can toggle layout containers on and off as described in Chapter 6, there is a way to change chart types with the click of a button. This second approach works best when you need to change between only two chart types, such as a visualization and its underlying table of data.

For this tutorial, we allow a user to toggle between a scatter plot and a bar chart, showing the underlying data for each dimension member in the scatter plot. To begin, place the primary sheet on a dashboard. For the best results, make the sheet floating so that you can control its exact location and dimensions. Here's how the scatter plot looks after adding it as a 500 × 500 pixels floating sheet:

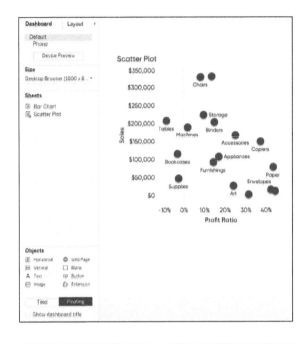

Next, drag a horizontal or vertical layout container onto the view. This object should be floating and have the exact same dimensions and x/y coordinates as the primary sheet. You can change these settings by going to the Layout pane on the left side of the interface:

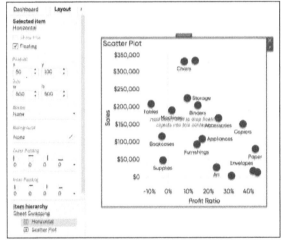

For the best results, make the background color of the layout container the same color as the rest of the dashboard. You can also find the background color settings on the Layout pane:

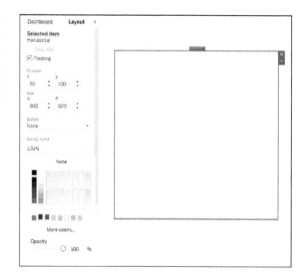

Now, place the secondary sheet that will be toggled on and off inside the layout container:

Your keen intuition is likely telling you that you can no longer see the scatter plot because the bar chart is covering it. Here is where the magic happens. As of Tableau 2019.2, you can toggle layout containers on and off. If we can turn off the container with our bar chart, we would again be able to see the scatter plot. If we can then toggle the container with the bar chart back on, we would be able to navigate our user back to the raw data for the scatter plot.

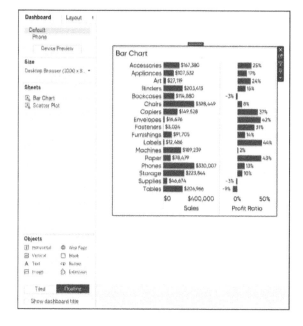

To toggle containers on and off, select the container and then, in the upper-right corner of the object, click the down arrow, and then choose Add Show/Hide Button:

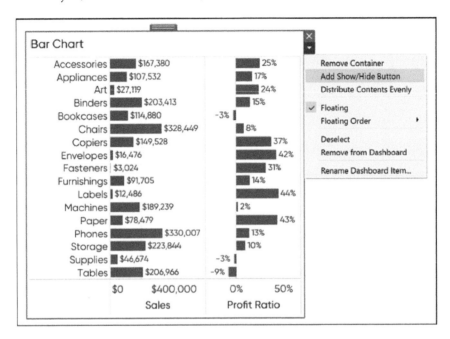

A new object appears with an image of an *X*. This is now a button that you can use to toggle the container with the bar chart on and off! You can test this experience by going to presentation mode; on the toolbar at the top of the window, click the projector icon and then click the button. By default, when the container is on, you will see the X to turn off the container; when the container is closed, you will see a hamburger menu to turn the container back on.

What's better is that Tableau allows you to customize the images used for both the toggled on and toggled off states of the container. This allows us to provide a slick user experience by, in this case, displaying a bar chart icon when the container is turned off, and a scatter plot icon when the container is turned on. This implies to the user that they can click the icon to swap in the respective chart.

To access these options, select the newly added Show/Hide button (outside of presentation mode), click the down arrow that becomes visible to access more options, and then choose Edit Button:

To select an image for the toggled-on state, in the Image subsection, click the Choose button:

To select an image for the toggled-off state, in the Button Appearance section, click Item Hidden, and then click the Choose button in the Image subsection:

Accept the settings by clicking OK.

This newly added Show/Hide button is its own floating object, so just like the scatter plot sheet and horizontal layout container we've added, we can change its precise size and location by modifying the settings on the Layout pane. Here's how the view looks after I resized the Show/Hide button and moved it more in line with my visualizations:

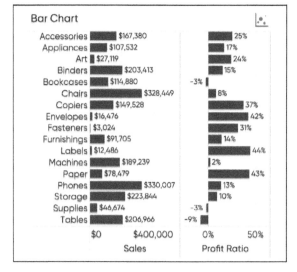

This view has the layout container turned on, and the scatter plot icon button will toggle the container off. After clicking the scatter plot button within presentation mode (or once this is published to Tableau Server or Tableau Public), the container is turned off, revealing the scatter plot that follows, and the toggle off button has now turned to the bar chart button to turn the container back on:

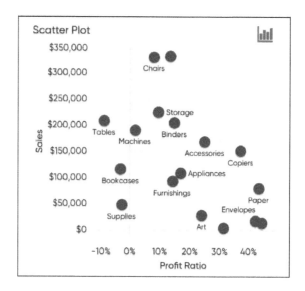

We just used a new approach to sheet swapping that allows the user to go back and forth between a primary scatter plot view and a bar chart illustrating underlying detail!

The tips outlined in this chapter are a natural segue into the next sections which cover tips for improving the authoring and user experience. Check out Chapter 70 for a tactic to make chart selection even better through the use of parameter actions. Instead of choosing a chart from a parameter control drop-down menu, as you've seen so far, you could design an experience that allows the user to choose the chart by clicking a dimension member on a dashboard.

PART III

Author Experience

CHAPTER 47

3 Innovative Ways to Use Text: Tip 1

One of the consistent challenges I face as a data visualization evangelist is to help my stakeholders evolve from viewing data in the Excel spreadsheets they have become accustomed to. Perhaps they want to do their own calculations on the data. Or, maybe they feel more comfortable seeing the exact numbers. Or worse, maybe "that's just the way they've always done it."

Beyond the *walls of numbers* I find myself battling against, I have come to realize that text really does play a critical role in the practice of data visualization. Although I strongly believe that visualization is the most effective way to understand data, text can complement your analyses, provide context, add branding, and much more.

So I thought what better way to begin Part III than to provide some ideas to help you get the most out of text in Tableau. Over the next three chapters, I share how to incorporate custom fonts, conditionally format text, align table text like traditional financial reporting, and more.

How to Install Custom Fonts

Even though Tableau comes out of the box with tens of dozens of fonts, you can add your own within both Tableau Desktop and Tableau Server. This is a very easy way to set your visualizations apart and/or create your Tableau dashboards within your company's brand guidelines.

To install custom fonts in Tableau, download and install them onto every computer at your company on which Tableau Desktop is installed. The next time you open Tableau Desktop, you will see the newly installed fonts under your text formatting options. If you are using Tableau Server, also install the fonts on the machine running Tableau Server to ensure that Tableau can recognize and render the appropriate fonts within published workbooks.

Note that if newly installed fonts are not rendering properly on Tableau Server, Tableau Online, or Tableau Public, it's not necessarily Tableau's fault. Tableau relies on different browsers to render text properly, and browsers don't recognize every font.

For this reason, it is best practice to pick web-safe fonts (*https://oreil.ly/QSbHG*) to better the chances that your end users all have a consistent experience.

If you must incorporate your brand's font into the dashboard and have it render the same for every user, one sneaky way to accomplish this is to type the text in a different program such as Word or PowerPoint, take a screenshot of the text, and add it as an image within the dashboard. It's difficult to make this text dynamic (i.e., change based on a selection), but it's useful in small doses such as adding custom titles to dashboards.

CHAPTER 48

3 Innovative Ways to Use Text: Tip 2

How to Conditionally Format Text

One of my favorite dashboard elements are large explanatory metrics that I call *callout numbers* (see Chapter 95 in *Practical Tableau*, O'Reilly, 2018). These are nothing but text tables built with oversized fonts that communicate KPIs at a very high level. I typically like to enhance these callouts with a comparison to either a goal or past performance. I show the callout number, followed by an up or down triangle, followed by the percentage or percentage point change. The tricky part is that I will color the triangle and change based on whether it was positive or negative.

Here's an example comparing a hypothetical current month sales to the prior month sales:

> SALES
> **$148,723** ▼ 8%

One way to achieve this is to have the text of the callout number on one sheet and the up/down triangle and percent change on a second sheet. You would then color the marks of the second sheet by whether the change was positive or negative. But I have a more elegant solution for you that allows you to conditionally format like this within a single sheet and provides the ability to conditionally format tooltips.

First, you must set up calculated fields for both positive changes and negative changes. I like to make my triangles and percent changes their own measures so that I can have different font sizes for each one. This means that I end up with four formulas: one for up triangle, down triangle, percent change when positive, and percent change when negative.

I use the IIF function within these calculated fields to display the triangle or percent change when the argument within the calculated fields is true; otherwise it displays NULL. We want to display NULL when the argument is false because we want only the positive or negative calculations to show up at a time; never both at the same time. Here are the formulas:

Up triangle
```
IIF([Report Month Sales]>[Comparison Month Sales],'▲',NULL)
```

Down triangle
```
IIF([Report Month Sales]<[Comparison Month Sales],'▼',NULL)
```

Percent change when positive
```
IIF([Report Month Sales]>[Comparison Month Sales],([Report Month
Sales]/[Comparison Month Sales])-1,NULL)
```

Percent change when negative
```
IIF([Report Month Sales]<[Comparison Month Sales],([Report Month
Sales]/[Comparison Month Sales])-1,NULL)
```

After you have the calculated fields, drag all of them to the Text Marks Card of the sheet with your callout number. Then, click the Text Marks Card to format the text. Put the elements in order (up triangle, down triangle, positive percent change, negative percent change), and format them to your liking. For this demonstration, I made the positive calculations blue and the negative calculations red:

Because of the way we set up our IIF calculations here, only the negative calculations display in red when the sales comparison is negative, and only the positive calculations display in blue when the sales comparison is positive!

Here's how the same callout number looks when the change is negative:

CHAPTER 49

3 Innovative Ways to Use Text: Tip 3

Format Text Tables Like Finance Reports

I am often asked to format text tables in Tableau with different levels of indentation for certain rows, like you might see in an accounting report. It's important to remember that Tableau does not make spreadsheets; even its crosstabs are *visual* representations of the underlying table of data. This is why you can't simply click into a cell and modify the value or the formatting for that cell.

If you need to have multiple alignments for your table headers, I have two simple tricks for you. For demonstration purposes, suppose that we want to indent the sub-categories in the Sample – Superstore dataset alphabetically: *A* sub-categories won't have any indentation; *B* sub-categories should be indented one tab (about five spaces); and *C* sub-categories should be indented two tabs (about 10 spaces).

I also filter out the other sub-categories to simplify my table. Imagine that the sub-categories are different subtotals or callouts within your finance table. Here's what the table looks like by default:

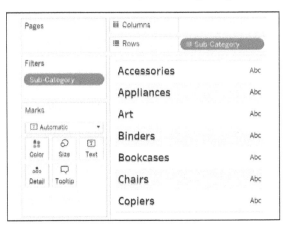

The trick to changing the alignment is to simply right-click each dimension member, and then, on the menu that opens, choose Edit Alias and add your leading spaces:

Note that this changes the sort order; spaces come before letters when sorting in alphabetical order. You can update this by clicking each dimension member and dragging it to the proper row. Here's how my table looks after changing all of the aliases and reordering my dimension members to get my desired alignment:

If you're worried about these new aliases showing up on other views within your workbook, you can duplicate dimensions by right-clicking them in the Dimensions area of the Data pane and then choosing Duplicate. This gives you two copies of the same dimension, and now you can change the aliases of one but not the other!

The second approach to aligning headers is to add the leading spaces in the underlying dataset.

Here's how an underlying table could look in Excel:

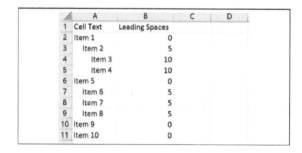

When you create the table in Tableau, the dimension members including the leading zeroes used to align them will be in the table.

Regardless of which approach you choose to implement, this flexibility allows you to match the text formatting of any report you have around the office—even the profit and loss statement!

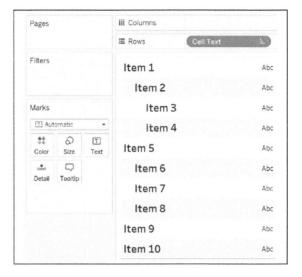

CHAPTER 50

How to Deduplicate Joined Rows

There are always problem-solving brainstorms at our Tableau training events (*https://oreil.ly/hF708*), and sometimes the solutions are so relevant for all Tableau users that I want to share them on a larger scale. In one such case, an attendee was trying to solve the business problem of currency conversion. This person worked for a global company and needed to join a dataset containing monthly exchange rates to their primary data source.

Joins in Tableau are a powerful way to add new dimensions and measures to your analysis, but without a good understanding of how they affect your dataset, you will often end up with inflated numbers. This chapter shares the challenge with joining multiple data sources and several solutions to ensure that you are getting accurate answers—even when joining on multiple dimensions (e.g., country and month).

The Challenge of Working with Joined Data Sources in Tableau

My favorite way to use joins in Tableau is to add fields to my analysis when my primary table and the table containing the new fields have at least one dimension in common. For example, suppose that I want to add a column from the Returns table to the Orders table of the Sample – Superstore dataset. The join would look like this in Tableau Desktop:

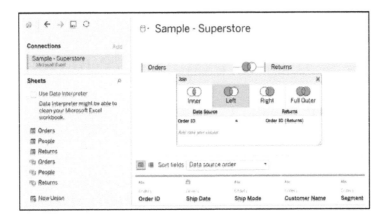

Because both tables contain the Order ID dimension, left joining the Returns table to the Orders table will add the new Returned column to the dataset. For each Order ID, I will have a Yes or Null to indicate which orders were returned. I could use this new column as a flag and/or create a calculated field to count the number of returns.

The problem is the Returns table tells me only whether an order was returned at the Order ID level, whereas the Orders table can have multiple rows per Order ID (in the case that multiple products were purchased in a single order). When this is the case, the join causes duplicate rows. For example, Order ID CA-2016-143336 contains three products, and the join creates three rows (two of them new) for each product:

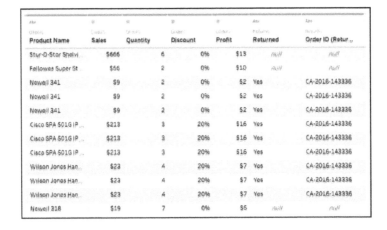

This can be very problematic for an untrained analyst because sales for the Newell 341 product in this order just went from $9 to $27.

How to Deduplicate Joined Rows Using Tableau's LOD Expressions

The easiest way for Tableau authors to correct this type of row duplication is to create a calculated field that uses a FIXED LOD expression to isolate one value for the most granular level of detail in the join. First, to illustrate how to identify an issue, let's look at one of the first things I do when working with a new data source in Tableau, which is to view the number of records.

You can do this by simply placing the generated field called Number of Records onto the Text Marks Card. In the default (i.e., unjoined) Orders table, this would result in 9,994 records:

However, if I do the same check for the joined dataset, I get 12,420 records. That means I have 2,426 duplicate records.

This becomes problematic when we try to use this joined dataset in a real-life scenario. Suppose I want to count the number of returned orders and have created this calculated field to tally the number one every time the Returned dimension equals Yes:

```
IIF([Returned]="Yes",1,0)
```

Placing this calculation on the Text Marks Card reveals the inflated answer of 3,226:

I know this number is inflated because the entire Returns tab in the Sample – Superstore dataset Excel file contains only eight hundred rows. If I want to count only one return per order, I need to deduplicate the Yes tallies at the Order ID level of detail.

If you use an aggregation of MIN, you will get one tally per Order ID:

```
{FIXED [Order ID]: MIN([Returns])}
```

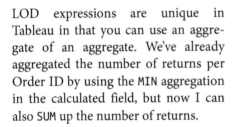

LOD expressions are unique in Tableau in that you can use an aggregate of an aggregate. We've already aggregated the number of returns per Order ID by using the MIN aggregation in the calculated field, but now I can also SUM up the number of returns.

This means that there was at least one item returned in 296 distinct orders.

How to Deduplicate Using a Row ID Field

The method described to this point is useful any time you want to use measures from the *secondary* data source, but you can do something similar to ensure that you are always using deduplicated values in the *primary* data source.

Suppose that we want to use the Sales measure from the primary data source; the unjoined sum of sales is $2,297,201:

Unfortunately, after joining the Returns table to the Orders table, we get the inflated answer of $2,901,677:

It is easiest to deduplicate rows in the primary data source when you have a dimension with unique Row IDs for each row in the data source. Little known fact: the Sample – Superstore dataset does have such a dimension, which you can access by clicking the down arrow in the upper-right corner of the Dimensions area of the Data pane and then choosing Show Hidden Fields:

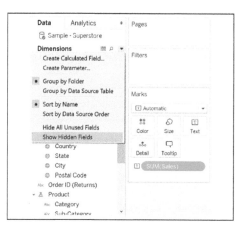

A dimension called Row ID will appear in gray lettering. After right-clicking the Row ID dimension and choosing Unhide, you can use the dimension in a calculated field:

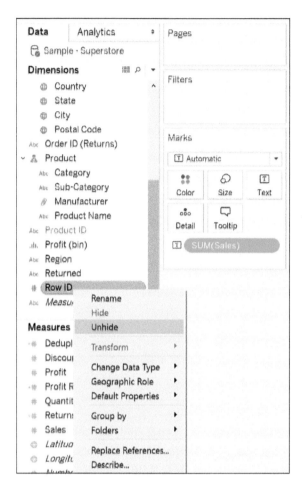

The formula for Deduplicated Sales is:

```
{FIXED [Row ID]:
MIN([Sales])}
```

This newly Deduplicated Sales results in the expected sum of $2,297,201:

How to Deduplicate Using LODs with More Than One Dimension

The global company trying to convert currency at the beginning of this chapter had an exchange rate had an exchange rate to US dollars (USD) for two dimensions: Month and Currency. To simulate the scenario, I've set up this Excel file containing one year of exchange rates for Canadian dollars and British pounds.

Because the Sample – Superstore dataset contains only USD, I'm going to pretend that we are localizing currencies for analysts in our Canadian and English territories (i.e., show what USD would be in their respective local currencies):

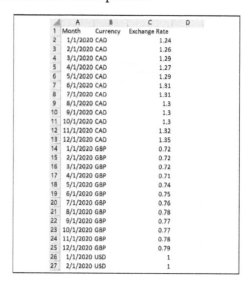

First, we left-join this exchange rate table to the Sample – Superstore dataset. Fields with a data type of Date have a lot of nuances, so to illustrate, I have added a column to the Sample – Superstore dataset to truncate every date at the first of the month:

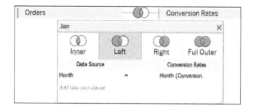

If I put a quick table together to look at 2020 exchange rates to make sure it matches the underlying Excel sheet, I once again see that the join has caused inflated values:

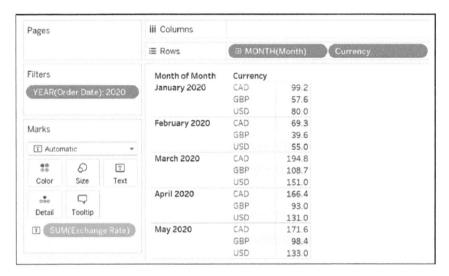

If I were to use this as is, my result would be way off and we might make an inaccurate insight that leads to wasted resources in our business.

To correct the duplicate rows and inflated values caused by the join, let's use a similar LOD expression that we used on the Returns table to deduplicate. The difference is that this time we have multiple dimensions to deduplicate. To deduplicate when there is more than one dimension, simply separate all the dimensions with commas immediately after the FIXED LOD.

In the case of our currency conversion table, the calculation looks like this:

```
{FIXED [Month],[Currency]:
MIN([Exchange Rate])}
```

After adding this calculation to the quality assurance table, we see that things are checking out:

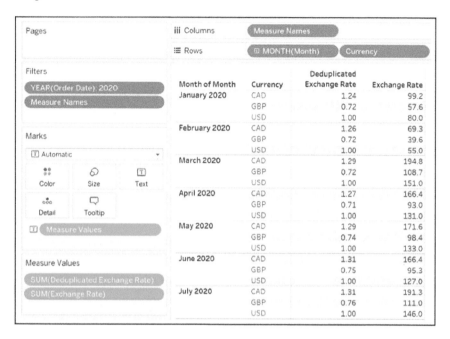

Pages			
iii Columns	Measure Names		
≣ Rows	MONTH(Month)	Currency	

Month of Month	Currency	Deduplicated Exchange Rate	Exchange Rate
January 2020	CAD	1.24	99.2
	GBP	0.72	57.6
	USD	1.00	80.0
February 2020	CAD	1.26	69.3
	GBP	0.72	39.6
	USD	1.00	55.0
March 2020	CAD	1.29	194.8
	GBP	0.72	108.7
	USD	1.00	151.0
April 2020	CAD	1.27	166.4
	GBP	0.71	93.0
	USD	1.00	131.0
May 2020	CAD	1.29	171.6
	GBP	0.74	98.4
	USD	1.00	133.0
June 2020	CAD	1.31	166.4
	GBP	0.75	95.3
	USD	1.00	127.0
July 2020	CAD	1.31	191.3
	GBP	0.76	111.0
	USD	1.00	146.0

Filters: YEAR(Order Date): 2020 / Measure Names

Marks: Automatic — Color, Size, Text, Detail, Tooltip, Measure Values

Measure Values: SUM(Deduplicated Exchange Rate) / SUM(Exchange Rate)

Now that I trust the exchange rate results, I can use this deduplicated calculated field in another calculated field to multiply our original monthly Sales values by the respective monthly exchange rate in each country.

The calculation would be something like this:

```
SUM([Sales]) * SUM(
[Deduplicated Exchange Rate])
```

Sales * Exchange Rate ⊗

```
SUM([Sales]) * SUM([Deduplicated Exchange Rate])
```

The calculation is valid.   Apply   OK

Here's a line graph with the USD sales values on the first row, and our converted sales values on the second row. On the second row, changing the Currency filter multiplies each month's USD sales value by the appropriate monthly exchange rate to correctly localize the amounts:

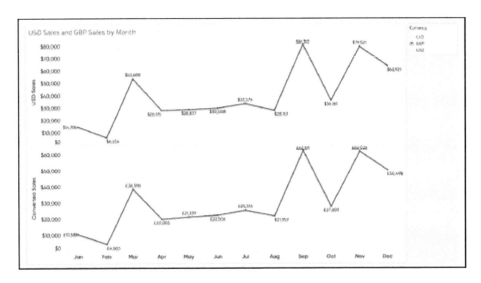

 Related: *Practical Tableau*, Chapter 65, "How to Dynamically Format Numbers" (O'Reilly, 2018)

I should note that in this last example, because we are using both the Month and Currency dimensions, I could have skipped the calculated field with the LOD expression and got the correct answer by manually changing the aggregation of the Exchange Rate measure to Minimum. However, deduplicating using LOD expressions has two benefits: you don't need to constantly worry about matching the correct aggregation with the visualization's level of detail, and you are allowed an extra aggregation; we did this earlier when we aggregated an aggregate by summing up minimum values.

For one more real-world application of using level of detail expressions to deduplicate joined rows—and just for a cool trick—see Chapter 82, "How to Turn Data Normalization On and Off."

CHAPTER 51

How to Pass Parameters and Filters Between Workbooks

As Chapter 75 demonstrates, I often try to separate large workbooks into smaller individual files. This helps with efficiency, managing fields, and maintaining focus on individual business questions. My favorite tactic for linking the individual workbooks on Tableau Public, Tableau Online, or Tableau Server is to provide a recognizable menu to my end users that contains cross-workbook links.

But did you know that not only can you provide the links in Tableau, you can pass filters and parameters *between* workbooks. For example, if you have parameters for a start date and an end date in Workbook 1, you can set up your menu link to pass the current values of those parameters to Workbook 2. This provides a seamless experience and helps you to avoid confusion related to how the destination workbook is being filtered. This chapter shows you how to pass filters and parameters between Tableau workbooks.

How to Pass Fields Between Workbooks Using Query Strings

To illustrate how to pass filters and parameters between workbooks, let's begin by changing the Minimum Date and Maximum Date parameters in a workbook that we'll call the origin workbook. Then, we link from there to the *Super Sample Superstore* visualization on Tableau Public (*https://oreil.ly/6Unk0*) to show you how we can change the start date and end date of a destination workbook.

To show you the current parameter settings, we can navigate to any sheet in the origin dashboard and show the parameter controls for Minimum Date and Maximum Date:

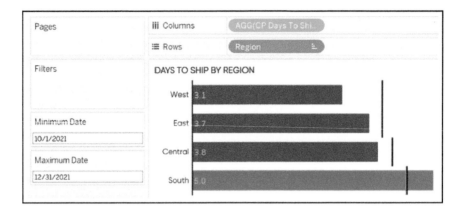

As you can see, the date range is set to 10/1/2021 to 12/31/2021. These are the same default settings of the *Super Sample Superstore* workbook. We're eventually going to overwrite the date range of the destination workbook on Tableau Public by passing new date parameters from the origin workbook. To begin, let's change the date range to 4/1/2022 to 6/30/2022:

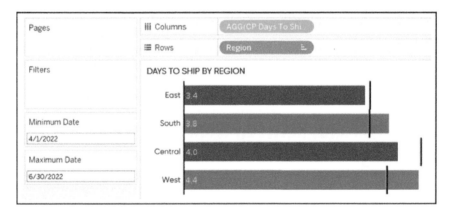

The trick to passing filters and parameters between Tableau workbooks involves dashboard actions and modifications to the URL string. To demonstrate this, let's go to a dashboard in the origin workbook and set up a URL dashboard action (Dashboard > Actions > Add Action > Go to URL in the top navigation):

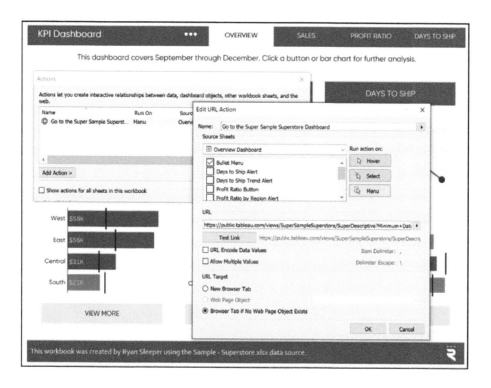

The URL for the destination workbook (which in this case is a Tableau Public workbook) looks like this:

```
https://public.tableau.com/views/SuperSampleSuperstore/SuperDescriptive
```

To pass the Minimum Date and Maximum Date parameter values to the destination workbook on Tableau Public, I am going to add this code to the end of the current URL:

```
?Minimum+Date=<Parameters.Minimum Date>&Maximum+Date=<Parameters.Maximum Date>
```

The syntax is simply [The name of the field you are changing at the destination] = [The parameter you are passing from the origin], preceded by *parameters*.

Here are a few things to note:

- The field from the origin should be surrounded by <> brackets.
- Add an & symbol between each new item you are passing.
- The first item should start with a ?, but if there is already a question mark symbol in the URL string, use an ampersand instead (the question mark should be used only once per URL).

- I added plus signs (+) to replace blank spaces in the URL.

The full URL now looks like this:

```
https://public.tableau.com/views/SuperSampleSuperstore/SuperDescriptive?Minimum
+Date=<Parameters.Minimum Date>&Maximum+Date=<Parameters.Maximum Date>
```

Now, if I click the link in the bullet menu, not only does the *Super Sample Superstore* workbook open in a new window, but it is filtered to the same parameter values as the origin. Note that the date range of 4/1/2022 to 6/30/2022 overwrote the original 10/1/2021 to 12/31/2021 range:

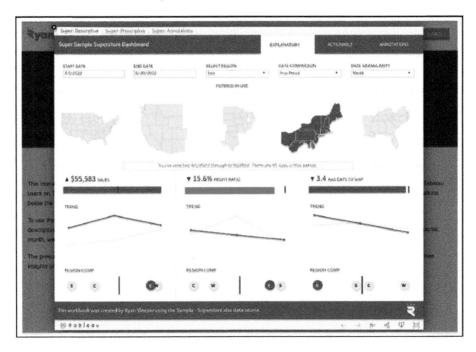

Passing filters works the same way as passing parameters. You can pass an individual dimension member (i.e., dimension filter) along with parameters. For example, perhaps you would like to pass not only a date range between workbooks, but also have the destination filter to a specific region. In this case, you would add a new URL dashboard action to the charts that contain the Region dimension and add this query string to the end of the URL:

```
?Region=<Region>&Minimum+Date=<Parameters.Minimum
Date>&Maximum+Date=<Parameters.Maximum Date>
```

The dashboard action looks like this:

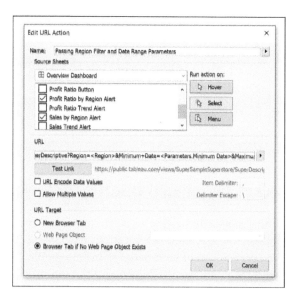

Now if I select East in one of the charts that are broken down by region, not only does the destination workbook open with the date range from the origin, but it also filters to the East region:

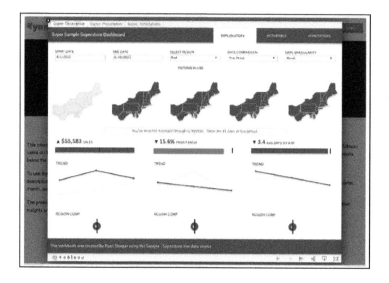

This example was purely for illustration purposes. You can pass whatever parameters, filters, or dimension members you want between workbooks, as long as both workbooks contain the same fields. This works on Tableau Public, Tableau Online, or Tableau Server.

CHAPTER 52

3 Innovative Ways to Use Dates: Tip 1

How to Equalize Dates Using the MAKEDATE Function

Comparisons and viewing trends across dates are two effective ways to turn raw data into insights, but dates can be tricky to work with in Tableau. In Chapter 26, I illustrate how to find hidden patterns in line graphs by adding a slope graph toggle, but what if the dates are not lined up on the same axis? For example, if you were to make a Sales by continuous Order Date line graph with Tableau's Sample – Superstore dataset and color the marks by year, you would get four colored lines that do no not line up on top of one another.

Tableau does not have a Month + Day date part, which can make it challenging to compare year-over-year performance. This chapter shows you a basic way to normalize months and days so that they share the same axis when colored by year. When the marks line up, it is much easier to evaluate year-over-year performance.

The Challenge of Not Having a Month + Day Date Part in Tableau

First, let's have a look at why this tip is needed. Here is an illustration of the sales by continuous order date line graph:

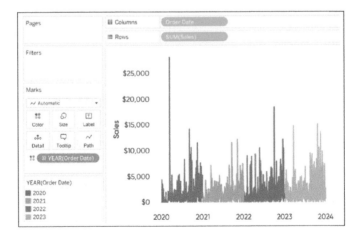

As you can see, although all four years are represented on the line graph, it is difficult to compare the same time period year over year. For example, it is almost impossible to determine whether the peaks in each year line up on or around the same dates. This would be much easier to evaluate if the lines for all four years were lined up on top of one another on the same axis.

Before I discovered the tip that I'm about to share, I would go through the manual process of adding the appropriate number of days to each historical year in my dataset to get them to line up with the current year; something like this:

```
IF YEAR([Order Date]) = 2023 THEN [Order Date]
ELSEIF YEAR([Order Date]) = 2022 THEN [Order Date] + 365
ELSEIF YEAR([Order Date]) = 2021 THEN [Order Date] + 365 + 365
ELSE NULL
END
```

This approach works, but has a few limitations:

- It requires the user to do the math to properly normalize year over year dates.
- It does not account for Leap Day.
- It is not scalable in that if more years are added, the formula doesn't automatically account for new dates.

Equalizing Year-Over-Year Dates on the Same Axis Using the MAKEDATE Function

An easier way to normalize year-over-year dates is to create a calculated field that combines the Month and Day date parts of each date in your dataset with the current year. We can achieve this by using the MAKEDATE function and this formula:

```
MAKEDATE([Current Year],MONTH([Order Date]),DAY([Order Date]))
```

Note that this formula is to normalize dates so that they are all in the year 2023. You would replace 2023 with the current year.

Now if I replace the original Order Date field on the Columns Shelf with my newly created Equalized Date field, the lines for all four years will line up in 2023:

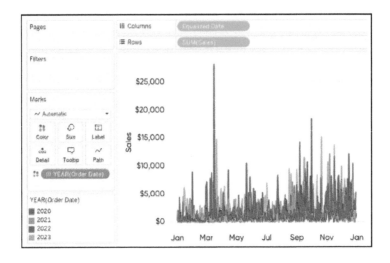

Admittedly, this view has too many marks to see much more than the spikes, but you can now filter on the Equalized Date field to display fewer marks. You can also change the date part of the Equalized Date field and everything will still line up year over year.

Here's how the view looks when changing the date part from Day to Quarter:

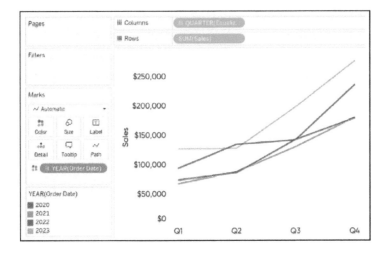

You can even take this a step further by allowing you and your end users to change the date part of the equalized date on the fly as described in Chapter 70; just replace Order Date with Equalized Date in the tutorial!

CHAPTER 53

3 Innovative Ways to Use Dates: Tip 2

How to Normalize Current Dates and Prior Dates on One Axis

One of my favorite techniques that doesn't happen out of the box for us in Tableau is to compare the performance of a selected date range to the performance of the date range immediately preceding it. For example, if I choose this week as the current date range, I want to see this week's data in addition to last week's data so that I can do an easy period-over-period analysis.

You can always set your date range filter to capture both time periods, but as Chapter 52 demonstrates, if you're using a continuous axis, the dates will not be lined up on top of one another.

This chapter shows you how to compare any selected date range to the same number of days immediately preceding the selection—all on one axis so they're lined up! Although I wrote about this in *Practical Tableau*, I found an even better approach to reduce the number of calculations required.

How to Equalize a Selected Date Range with a Prior Date Range on One Axis in Tableau

Before I share the technique for normalizing dates so that they line up period over period, let's take a look at why this is needed.

Suppose that we want to look at the sales performance for the month of October 2023. No problem; here's the Sales measure by continuous Order Date from the Sample – Superstore dataset, filtered to the month of October 2023:

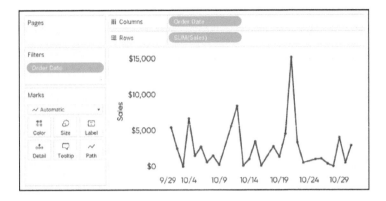

Now let's say that we want to do a month-over-month comparison. I'll bring in the September dates by changing my date filter to start at September 1:

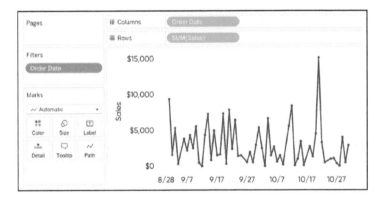

Even though both the current selection and comparison period are represented on the view, it's very challenging to analyze the month-over-month performance because they aren't directly lined up on the graph. For example, the spike you see is the date of October 21; if you want to see how we did on September 21, you would need to scroll over the marks and read the tooltips to find the same date last month. Even then, it would be challenging to compare the scales because the data points aren't lined up.

You could change the Order Date dimension to discrete Day and then put MONTH([Order Date]) on the Color Marks Card to equalize the dates, but I'm going to show you an approach that has several advantages:

- The comparison date range will automatically be computed for you.
- You can use a continuous axis, which is my preference with line graphs.
- You will see a line with all of the *dots* connected (discrete headers would show NULL by default when there is no data for a date).

- We can use the approach regardless of which date part you're using to truncate the order date.

- It creates a true apples-to-apples date range comparison. For example, if you select 31 days in October, the comparison will be the 31 days immediately preceding October; so, the 30 days in September plus one additional day to make the ranges the same number of days.

To begin, create two parameters with a data type of Date; one parameter will be used to control the beginning of the date range, and the other will be used to control the end of the date range.

For the Start Date parameter, let's set the current value to the first of October and leave the Allowable values as All:

For the End Date parameter, we set the current value at October 31 and, again, we leave the Allowable values set to All:

Next, we need a calculated field that computes the number of days in the range selected between the Start Date and End Date. We use

this later on to equalize the dates from the comparison period so that they line up with the dates of the current period.

Here's the formula:

```
DATEDIFF('day',[Start Date],
[End Date])+1
```

The next calculation we need will determine whether each date is in our current date range or our comparison date range. Here's the formula for that:

```
IF [Order Date] >= [Start Date] AND [Order Date] <= [End Date] THEN "Current"
ELSEIF [Order Date] >= [Start Date] - [Days in Range] AND
[Order Date] <= [End Date] - [Days in Range] THEN "Comparison"
ELSE "Not in Range"
END
```

When the dates are within the Start Date and End Date parameter values, this calculation classifies those dates as the current range. To compute which dates are in the comparison period, the calculation takes the values set in the two parameters minus the days in the range, creating a date range with the same number of days immediately preceding the selected range! The ELSE statement at the end acts as a catchall that will classify dates that did not meet either of the first two statements; we filter these out later anyway.

The last calculated field we need adds the number of days in the range to the comparison date range so that they line up on the same axis with the selected range. I call this the Date Equalizer:

```
IF [Current / Comparison] = "Comparison" THEN [Order Date] + [Days in Range]
ELSE [Order Date]
END
```

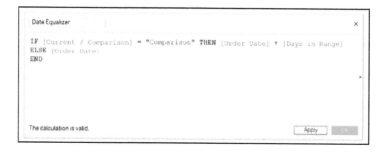

We're now ready to make the chart.

To illustrate what our calculations are doing, let's start this new version of the line graph by placing the Current / Comparison calculated field on the Color Marks Card and show the parameter controls for both the Start Date and End Date parameters. As you can see, the dates within the current selection are one color, and the dates in the range immediately preceding the selection are colored a second hue:

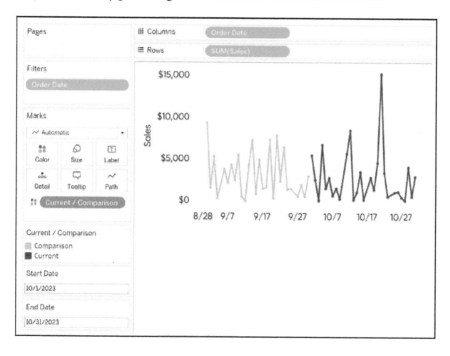

And here's exactly what I was talking about regarding why this makes it challenging to compare period over period dates. It's nice to see the different colors for each date range, but they are not lined up on the x-axis.

One of the interesting aspects of this approach is that we no longer need a date filter. The Current / Comparison calculated field is now determining which dates are left on the view, so let's remove the original Order Date filter from the Filters Shelf:

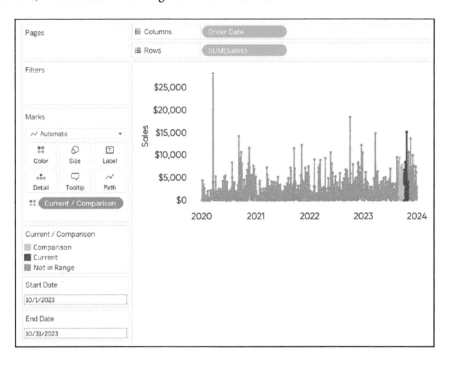

Now, we temporarily see all the dates in the dataset, with only our Current and Comparison date ranges colored. We can filter out the dates that are not in range by placing the Current / Comparison calculated field on the Filters Shelf and excluding Not in Range:

Lastly, to move the comparison dates forward on the x-axis so that they align with the dates in the selected range, we need to use the newly created Date Equalizer calculated field instead of the original Order Date field. Here's how the chart looks after replacing Order Date on the Columns Shelf with the Date Equalizer field:

Now the comparison date range has not only been computed, but the number of days in the selected range has been added to the comparison range. This moves the comparison points forward so that they line up perfectly on the same axis with the selected range. Plus, the comparison date range is the same number of days as the selected date range. Most companies would look at October compared to September, but they're comparing 31 days to 30 days. In this view we're seeing all of October compared to September plus the last day in August to make the date ranges equal.

This approach even works with different date parts and regardless of what date range is selected. Here's the same view after changing the date part to continuous week and changing the selected range to Q4 2023 (through 12/29 to filter out partial weeks):

We've just created a quarter-over-quarter comparison, and these calculated fields have equalized the dates, allowing us to easily compare the same weeks quarter over quarter on one axis!

CHAPTER 54

3 Innovative Ways to Use Dates: Tip 3

How to Make Better Relative Date Filters

Have you ever needed to filter to the last 7 days, 30 days, 90 days, and so on in Tableau? You've likely found that it's easy to add a *relative* date filter by dragging a dimension with a data type of Date to the Filters Shelf, choosing Relative Date, and then showing the filter to access the date range options. These default relative date filters are *OK*, but they have limitations, including:

- Limited and static anchoring options
- No formatting options
- Requiring manual range selection

This chapter shows you how to make relative date filters in Tableau that overcome all of the limitations of the default relative date filters. You will be able to set up any ranges you want and have one-click access to change date ranges on the fly.

How to Create Flexible and User-Friendly Relative Date Filters in Tableau

The trick for creating relative date filters in Tableau that are flexible and more user friendly than the defaults involves creating *sets* and customizing each range on the Condition tab of each set. The Condition tab within a set allows you to set up dynamic date ranges that automatically update. By the end of this chapter, you'll know how to set up comparison date ranges and allow users to choose which date range is being used.

Let's say that I am writing this tutorial on July 14, 2023, and our first relative date range will be the last full seven days. Assuming that our dataset updates only once per day overnight (so our newest date is 7/13/2023), the condition would be dates greater than or equal to today minus seven days.

To create this set of dates, in the Dimensions area of the Data pane, right-click your date field, hover over Create, click Set, and then navigate to the Condition tab. Here's how the settings look for this first range using the Order Date field in the Sample – Superstore dataset (which I updated to have 2023 dates):

Note that this approach works only when the data type is Date (as opposed to Date *and Time*). The 7 in our logic represents the most granular date part, which in this case is *day*.

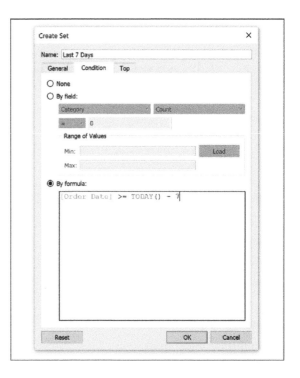

Let's quality check that the correct date range is being computed by creating a sheet with continuous Day of Order Date on the Columns Shelf and the newly created Last 7 Days set on the Filters Shelf. Sets are Boolean, so each Order Date is either in the set or not, and when a set is placed on the Filters Shelf, the default behavior is for Tableau to display only the dimension members *in* the set:

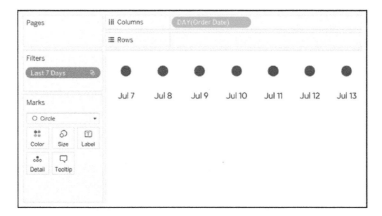

So far so good—we have seven days in the range, and they're the last seven days for which we have data.

This type of relative date range that we created by using the Condition tab of a set we can use for any range of dates. To finish the illustration, let's create two additional relative date sets: one for the last 30 days and one for the last 90 days:

Relative Date set keeping the last 30 days

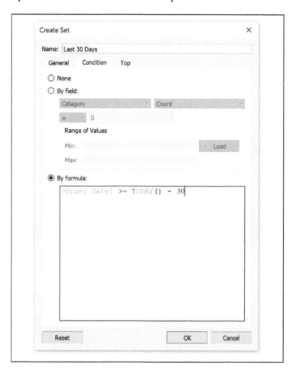

Relative Date set keeping the last 90 days

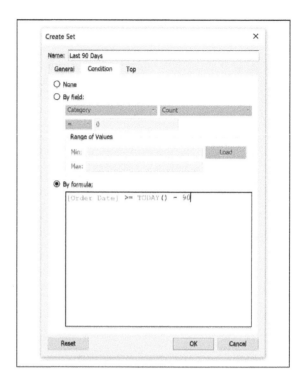

How to Allow Users to Select the Relative Date Range

Now that we have our relative date range options set up, we can allow users to decide which set is filtering the dates on the view. This works the same way as the classic Tableau tactic of allowing users to choose which dimensions and measures are being used in a view.

First, create a parameter with a data type of String and your list of options:

Next, create a calculated field using CASE/WHEN logic to instruct Tableau which set to use when each value within the parameter is selected. Note that the outcomes on each line are the sets created in the previous section of this chapter:

```
CASE [Relative Date Range]
WHEN "Last 7 Days" THEN [Last 7 Days]
WHEN "Last 30 Days" THEN [Last 30 Days]
WHEN "Last 90 Days" THEN [Last 90 Days]
END
```

Related: *Practical Tableau*, Chapter 64, "Allow Users to Choose Measures and Dimensions" (O'Reilly, 2018)

Let's try out this newly created filter on this sales-by-day line graph built with Sample – Superstore data:

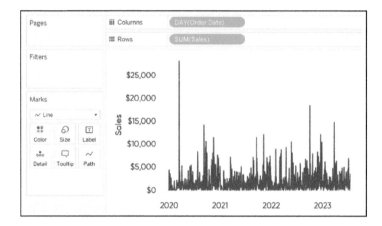

The three and a half years of daily data are making this view very challenging to analyze. If I place the Relative Date Filter calculated field on the Filters Shelf and choose True, we will see only the last seven days kept on the view. This is because the current value of the parameter that is feeding the calculated field is Last 7 Days, and when that is the selection, we are filtering on the Last 7 Days set:

If I show the parameter control for the Relative Date Range parameter—which you can do by right-clicking the parameter and then selecting Show Parameter Control—the user can choose Last 7 Days, Last 30 Days, or Last 90 Days. Here's how the view looks after updating it to Last 30 Days:

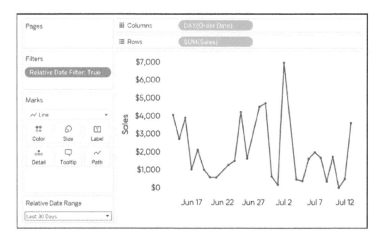

Combine This Relative Date Filtering with Parameter Actions for a Better User Experience

Because we are using a parameter to filter on our relative date sets, we can use parameter actions to allow our user to change the relative date range with the click of a button!

Set up an Excel file with your choices.

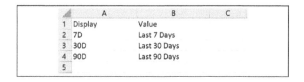

The Value field in these types of data sources that are used to create control sheets must match the values in the parameter. Make the Display field whatever text you want displayed on your navigation.

Next, connect to this Excel data source and create a text sheet that will act as our navigation. You can format this however you would like, but the Value field must be somewhere on the view for this to work during the last step:

Place both the control sheet and the chart(s) containing your relative date filter on a dashboard:

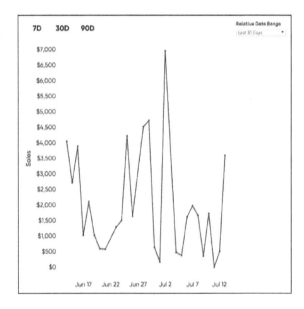

Lastly, create a parameter dashboard action; on the menu bar at the top of the window, click Dashboard > Actions. You will need to add a Change Parameter dashboard action. Here's how the settings look if we want the selection on the control sheet to populate the allowable value of the Relative Date Range parameter that is filtering the line graph:

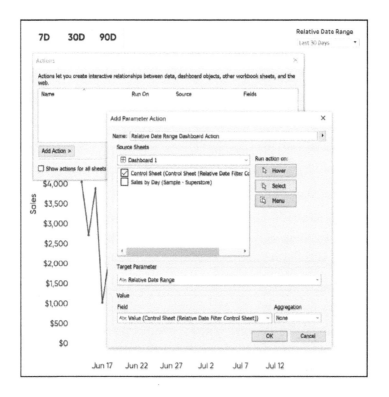

After accepting these settings, just clicking one of the three relative date range options will filter the line graph! Here's how it looks after clicking the 90D button:

Notice that the parameter action overwrote the allowable value in the parameter control. We no longer need the parameter control at all because the selection is now being controlled by the text sheet.

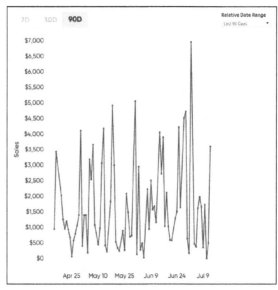

For more ideas on how to use this approach, see Chapters 69 through 71 at the end of Part III.

More Ways to Use These Better Relative Date Filters

We also can set up these relative date filters for comparison periods. For example, if you want to create a set of dates for the seven days immediately preceding our Last 7 Days set, the condition would be like this:

If you were to create a calculated field tied to the Relative Date Range parameter, like we did earlier in the Relative Date Filter calculated field, the current value of the parameter would instantly filter both your current relative date ranges and their respective comparison ranges. You can use these comparisons to create one of my favorite dashboard elements, the current versus comparison index callout (shown in Chapter 39).

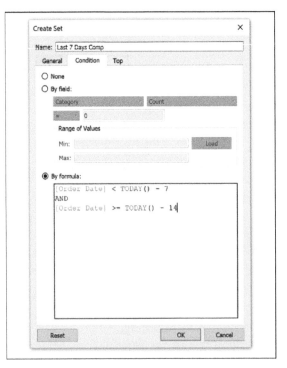

One more thought on how you can use these improved relative date filters. The code shared throughout this chapter uses today's date as the anchor. The TODAY() function will dynamically update the values in the set based on the current date. However, you also can use the DATE function to hardcode or even parameterize the anchor.

Suppose that you want to analyze the performance leading up to Christmas Day. Here's how the condition for the relative seven days set would be:

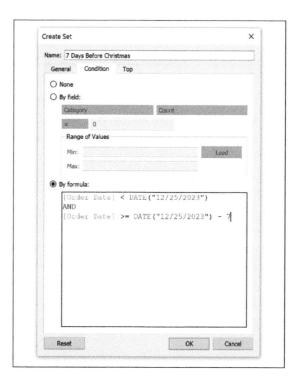

You could also replace DATE("12/25/2023") in this example with a date parameter that allows the user to choose any anchor they want:

We've just set up relative date filters with optional comparisons that allow users to choose the number of days being filtered and the anchor date of the filtering just by clicking options in a control sheet!

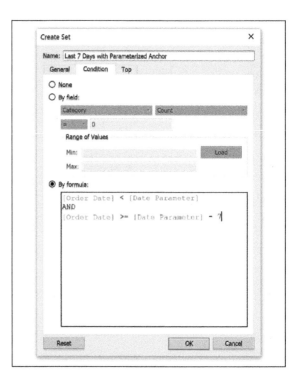

CHAPTER 55

3 Innovative Ways to Use Tooltips: Tip 1

How to Add an Image to a Tooltip

Chapter 9 discusses the long-awaited Tableau feature, Viz in Tooltip. I believe this will be a game-changer in how we blur the line between explanatory and actionable analytics. There's no doubt this feature has the ability to reduce the time to insight and increase the accuracy of insights; two of my three tenets of good visualization.

My third tenet of good visualization is to make them more engaging, and with infinite applications, the Viz in Tooltip feature can also help us draw users in, and even add some company branding. This chapter shares how to use Tableau's Viz in Tooltip to add an image to a tooltip.

How to Add Image Branding with the Viz in Tooltip Feature

For the past few years, I've used the following Tableau Public visualization (*https://oreil.ly/2opxL*) to share my speaking schedule:

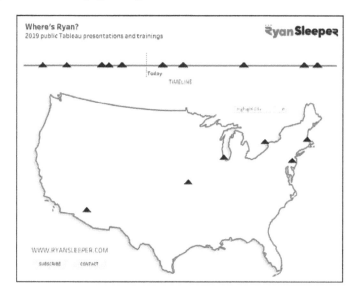

The Viz in Tooltip feature provides a great opportunity to integrate your own branding through the use of images. With the addition of this feature, I was able to improve the map visualization by adding my own custom images to the tooltips on hover. Now, when the viewer hovers over a date for one of our public Tableau training events, they see a product image along with the existing event information:

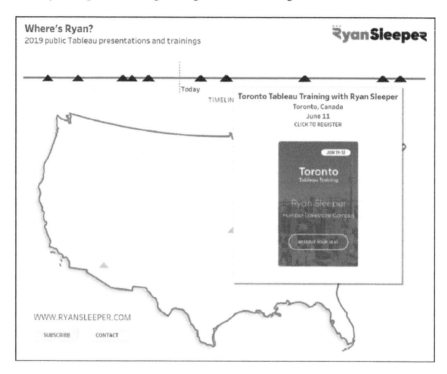

To add an image to a tooltip, begin by putting the images that will eventually be used into the Shapes folder in your Tableau Repository. On a PC, this is located at *C:\Users \User\Documents\My Tableau Repository\Shapes*. Here you can create a new folder for your custom shapes. You can name it anything you like—and create as many individual folders as you want—but for this exercise, I will place the images in a new folder called *Training Images*:

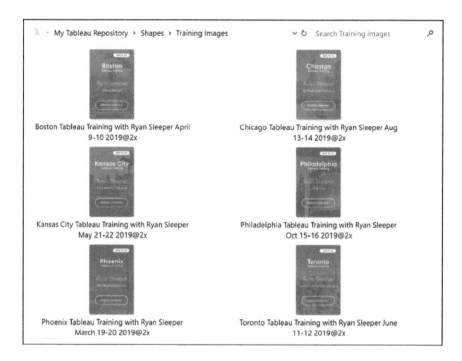

Boston Tableau Training with Ryan Sleeper April 9-10 2019@2x

Chicago Tableau Training with Ryan Sleeper Aug 13-14 2019@2x

Kansas City Tableau Training with Ryan Sleeper May 21-22 2019@2x

Philadelphia Tableau Training with Ryan Sleeper Oct 15-16 2019@2x

Phoenix Tableau Training with Ryan Sleeper March 19-20 2019@2x

Toronto Tableau Training with Ryan Sleeper June 11-12 2019@2x

Related: *Practical Tableau*, Chapter 47, "How to Create Icon-Based Navigation or Filters" (O'Reilly, 2018)

Next, make a new worksheet that will be used to map the new images to specific dimension members. This is like the approach used to create icon-based filters in my *Odds of Going Pro* viz (*https://oreil.ly/dZ4XV*). For best results, put the dimension to which you want to map the images on the Rows Shelf, change the mark type to Shape, and place the same dimension you are mapping onto the Shape Marks Card. After I filter my view to only the dimension members that I want to map, my sheet looks like this:

To map the images to dimension members, click the Shape Marks Card. Use the drop-down menu to find the custom shapes palette that you added to your Tableau Repository. Note that if you already had Tableau open, you will need to click Reload Shapes for the new palette to appear:

To map each image, click the dimension member first and then the shape that you want to assign to that dimension member. To preview the change, click Apply:

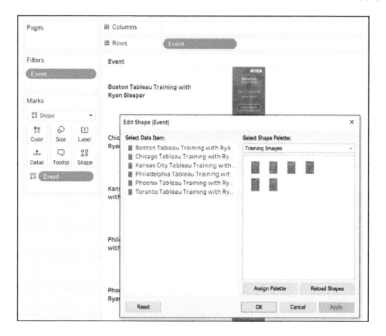

Click OK to accept the changes. Finalize the sheet by hiding the row headers (right-click any header and deselect Show Header), and remove all lines by using the formatting options:

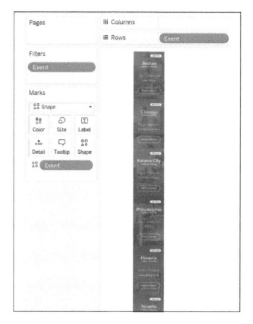

Lastly, navigate to the sheet where you want to add an image in a tooltip. To add a Viz in Tooltip, click the Tooltip Marks Card and then, toward the upper-right corner, click Insert, hover over Sheets, and choose the sheet containing your images:

Note that you can also set a maximum height and width of the image:

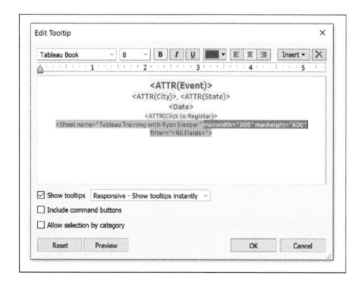

Now if the dimension member that is hovered over where you added the Viz in Tooltip matches the dimension member from the worksheet containing the images, the corresponding image will be displayed in the tooltip! I used the Event dimension, but this approach will work with whatever dimensions are relevant for your business.

CHAPTER 56

3 Innovative Ways to Use Tooltips: Tip 2

How to Add a Tooltip to a Dimension

Tableau tooltips are one of the best tactics for providing context to a data visualization without taking up valuable real estate on the visualization itself. By default, Tableau tooltips can be added only to measures, but what if you want to provide additional information about text on a row or column header?

The following trick allows you to display the definition of the dimension members on your view when a user hovers over them. This has several practical benefits, such as:

- It helps your end users understand the dimensions in the flow of the analysis.
- It's more intuitive to display information about the dimension on the dimensions themselves.
- It improves focus by saving real estate in the measure tooltips.

How to Display Definitions in Tooltips of Discrete Dimensions

Let's imagine that our business use case is to make the following bar chart showing the Sales and Profit Ratio measures by the Segment and Ship Mode dimensions:

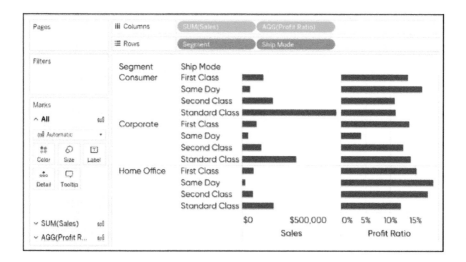

By default, if I hover over any of the bars, I see a tooltip:

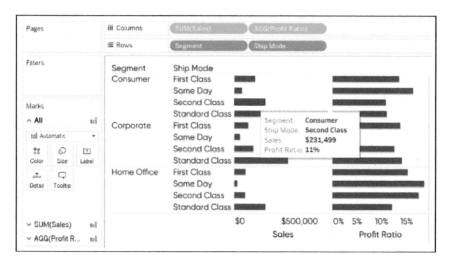

However, if I hover over a dimension member on one of the row headers, I see nothing:

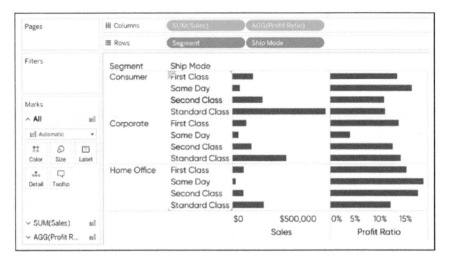

But what if we want to provide some information about our row headers? Where did they come from? How are they defined? Information about dimensions can be just as important to an analysis as information about measures.

Yes, I understand that you can add information about the dimensions to the tooltip for the measures, but the approach I'm about to show you has at least two big advantages:

- Information about dimensions can be placed in better relation to the dimensions themselves.

- Information about dimensions is typically more static than information about measures, so this approach allows you to save real estate on the more dynamic measure tooltips by moving information to the more static dimension tooltips.

To add a tooltip to a dimension, we once again use the placeholder hack.

To begin, create a calculated field for which the entire formula is as pictured:

Now, let's rebuild the bar chart that we looked at a moment ago, but instead of placing the Sales and Profit Ratio measures on the Columns Shelf, let's place this newly created Placeholder field on the Columns Shelf. Note that the dimensions on the Rows Shelf will stay the same, but the Placeholder field will be placed on the Columns Shelf four times; the first two will eventually be the dimensions to which we are adding a tooltip, and the second two will eventually be the Sales and Profit Ratio measures:

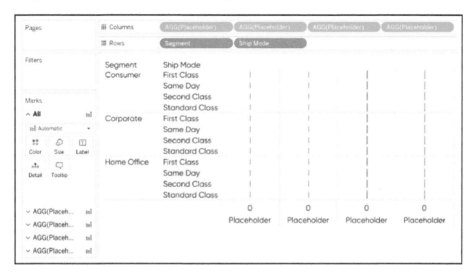

At this point, we have the framework for our bar chart. What's powerful about this second approach to creating the bar chart—and will provide the flexibility to add a tooltip to a dimension—is that each column now has its own set of Marks Cards. This means that we can independently edit them.

Let's begin re-creating our earlier bar chart by navigating to the third and fourth Marks Shelves, changing the mark type to Gantt Bar, and placing the Sales and Profit Ratio on the Size Marks Card for each respective Marks Shelf:

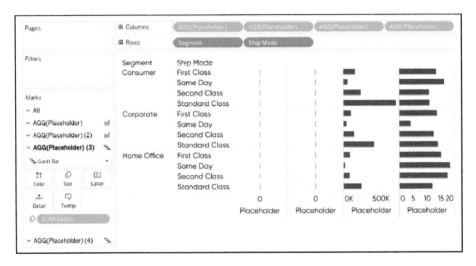

Here's where the trick for adding a tooltip to a dimension comes in. For the first two Marks Shelves, we change the Mark Type to Text and place the Segment and Ship Mode Dimensions on the Text Marks Card for each respective Marks Shelf:

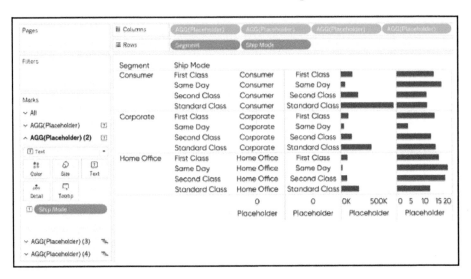

Note that because the dimensions are on the view as measures (because they are technically the text being displayed for the Placeholder *measure*), they have a complete set of Marks Cards. This means that you can now add a tooltip to a dimension by modifying the Tooltip Marks Card! We can use the tooltip to show the definition of the measure, what data source it comes from, whether the data source is verified, additional context about the visualization, or whatever other purpose you can think of:

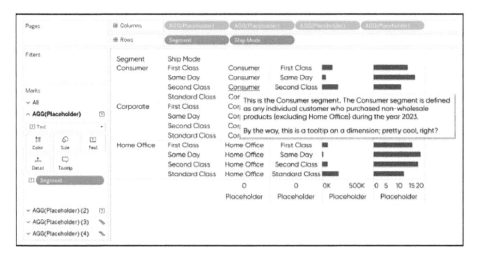

Let's finish the view by cleaning up the formatting, including the following:

- Hide the first two columns (the original dimensions on the Rows Shelf) by right-clicking them and deselecting Show Header.

- Remove the zero lines and gridlines on the text columns by right-clicking either column, choosing Format, navigating to the Lines tab, and then turning the zero lines and gridlines to None.

- Hide the Placeholder axes by right-clicking any of the axes and deselecting Show Header. Note that this is one drawback of the Placeholder approach; you need to float the column headers and/or axes on a dashboard if you want them. For the latter, directly labeling marks is a nice alternative.

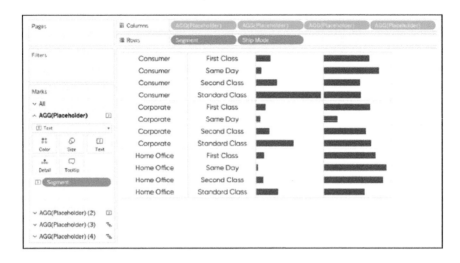

Lastly, if you're losing sleep because the segment names are repeated for each row in the first column, there's a hack for that, too. You can color the text on the first row and color the remaining rows white using a simple table calculation.

First, make a calculated field with the formula INDEX() = 1. Next, put the newly created First Row calculation on the Color Marks Card for the first column. Change the direction and scope of the table calculation for the First Row calculated field to Pane (down). Pick a color for when the statement, INDEX() = 1, is True and map white for when it is False:

We just re-created an entire chart type using a series of placeholder fields, and now we have the flexibility to customize the tooltips for every field—even when the fields are discrete dimensions!

CHAPTER 57

3 Innovative Ways to Use Tooltips: Tip 3

How to Make Dynamic Tooltips

Whenever I know that my data visualizations will be consumed via an interactive version of Tableau (i.e., Tableau Public, Tableau Online, Tableau Server, Tableau Reader), I move as much secondary information as possible to the tooltips. This is easily done by moving the fields you want displayed in the tooltips to the Tooltip Marks Card and formatting as desired.

However, there is one big drawback with tooltips. Any field placed on the Tooltip Marks Card will be shown for *every* mark on the view, and in the same format. There are times when this doesn't make sense, resulting in strange results such as descriptive words in the tooltip that don't have corresponding values (which happens if the value is null). This chapter shares how to make dynamic tooltips in Tableau, allowing you to show different information for each mark.

How to Automatically Display Varying Information Within Tooltips

To illustrate how to make dynamic tooltips in Tableau, we once again dig into *A Tale of 50 Cities*, which shows the census-over-census population change of the 50 largest cities in the United States since 1790. Notice there are no labels on the highlight table, so the information was communicated through the tooltips, which you can see if you hover over the interactive version (*https://oreil.ly/DUgjM*):

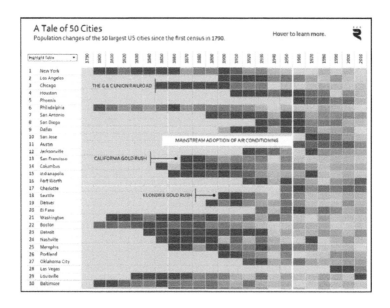

While creating the visualization, I realized there were two issues with my tooltips. First, I wanted to display the population for each city during each census, but my data source only had the population for cities that were in the top hundred at the time of each census. This prevented me from making a generic tooltip that said something like [Year of Census] Population because if the city were not in the top hundred, this descriptive text would be followed by a null value/blank space:

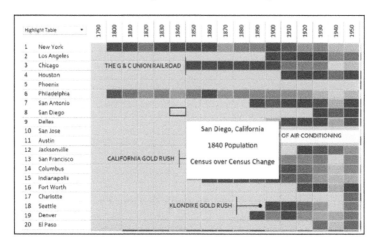

In the screenshot, you can also see the second issue. If there is no data for Tableau to compute the census-over-census change, the Census over Census Change label is also followed by a null value/blank space. This is a potentially confusing user experience that we want to avoid.

The trick to creating dynamic tooltips is to create a calculated field that provides logic for what descriptive text to display for each mark. In my first case, if a city did not have a Population measure because it wasn't in the top hundred, I would prefer that my end users see (Not in Top 100 Cities) instead of the standard text for the other marks followed by a null value. Here's the formula for this calculated field:

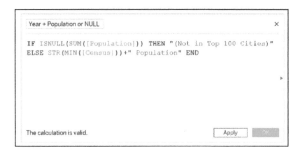

```
IF ISNULL(SUM([Population])) THEN "(Not in Top 100 Cities)"
ELSE STR(MIN([Census]))+" Population" END
```

This calculation looks at the population measure for each mark. If it is null, it displays (Not in Top 100 Cities); otherwise, it strings together the year of the census and the word Population.

This approach also works with table calculations.

Here's how I created a dynamic tooltip for the census-over-census change:

```
IF SUM([Census]) = 1790
THEN "First Year of US Census"
ELSEIF
ISNULL((ZN(SUM([Population])) - LOOKUP(ZN(SUM([Population])), -1)) /
ABS(LOOKUP(ZN(SUM([Population])), -1))) THEN "Census over Census Change N/A"
ELSE "Census over Census Change"
END
```

Note that this descriptive text for the tooltip has three outcomes:

- The first year of the census
- Census-over-census change is not available (as is the case when there aren't values for Tableau to do the math)
- The original standard Census over Census Change

After creating these calculated fields to dynamically compute the descriptive text that will be displayed in the tooltip, we simply replace the hardcoded text in the tooltip with these fields:

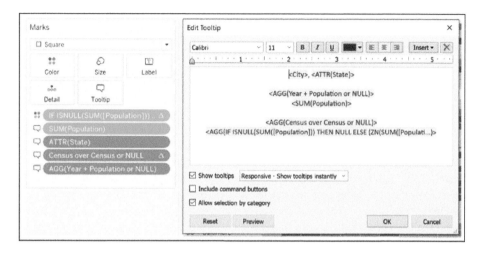

And now when an end user hovers over a city that was not in the top hundred at the time of the census and/or a city where the data was not available to compute the census-over-census change, they see a description that makes much more sense:

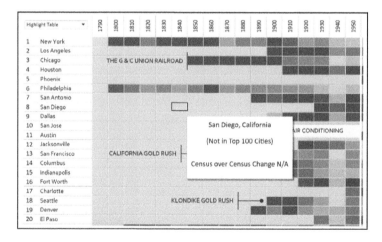

This is just one of countless examples, but the main point of the chapter is that you can take advantage of calculated fields with strings, the ISNULL function, and/or table calculations (plus more that we didn't cover) to improve the user experience with your tooltips!

CHAPTER 58

How to Use INDEX() for Easier Top N Filters

With each new field you add to the Filters Shelf in Tableau, you increase the complexity of the view and it becomes increasingly challenging to manage the combination of filters being used. Each filter acts as an AND statement, meaning that all criteria between every filter must be met for the mark to show on the view. To make things trickier, some of the filters can be set to include marks, whereas others can be set to exclude marks. To make things *even* trickier, you can have measure filters and dimension filters, but the Condition tab in a dimension filter can include measures—what?

Sometimes, you simply want to show the top *N* (e.g., top 5, top 10) for whatever is left after entering all the criteria. This chapter provides a very quick tip that I sometimes use to make my filters easier to manage and more predictable. This trick has the potential to not only improve the user experience of a view, but also of the authoring experience itself.

How to Create More Predictable, Flexible Top N Filters

To illustrate this tip, consider the following view showing Sales by Customer Name in the Sample – Superstore dataset. The view is currently filtered to show customers in the East region who have spent at least one thousand dollars:

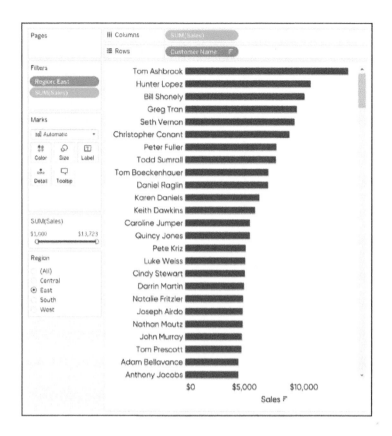

Now, suppose that you want to keep the top three names on the view: Tom, Hunter, and Bill.

Your first instinct might be to add a filter for Customer Name, navigate to the Top tab, and set it up to keep the top three by SUM(Sales). Also ensure the setting on the General tab is set to "Select from list" with all customer names included:

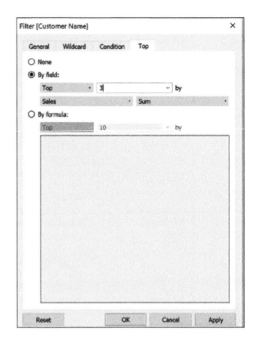

And we're left with Tom, Hunter, and…Ken?

The dimension filters are acting as AND statements because they are happening at the same time within the filter order of operations. Bill is not in the top three customers by SUM(Sales) overall; he is only in the top three of the East region. We could add the Region filter to context to get the result we are looking for, but that's a conversation for a different chapter.

Here's a trick I like to use for faster, more predictable results. Set up a calculated field that looks like this, replacing the 3 with the number of records that you want to keep:

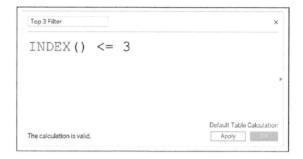

INDEX() <= 3

In this case, INDEX() is synonymous with row number, so if your view is sorted in ascending order, the top three will be kept on the view. If your view is in descending order, the first three rows will be kept, which is actually the bottom three performers (of whatever rows are left from your other filters).

The formula is binary, which is very efficient because there are only two outcomes to compute; the row is either less than or equal to three or it's not. You could have also written this formula as INDEX() < 4.

Here's how the view looks when I replace the Customer Name filter with this newly created Top 3 Filter:

As you can see, we're back to the expected result. What's great about this is the first three rows will always be kept, no matter which combination of filters is being used.

To improve the user experience, you could *parametertize* the number being used for the top *N*, allowing your end user to choose how many rows are being displayed. So, instead of hardcoding a number in the Top *N* Filter calculated field, replace it with a parameter that has allowable values from which your end user can choose.

CHAPTER 59

How to Dynamically Display the Top N Versus Other

There are times when I like to see every dimension member represented in a data visualization. For example, scatter plots are a great choice whenever you want to view many data points in the same space. I'll also occasionally create a bar chart or histogram with every dimension member, even if it causes a vertical or horizontal scroll bar to appear.

Although the *long-tail* insights these charts can provide have value, I tend to focus on the few dimension members causing the biggest impact on the business I'm analyzing. For this reason, I often show the Top *N* (usually Top 5 or Top 10) and group everything else into its own segment. And while I don't advocate the use of pie charts (see alternatives in Chapter 52 from *Practical Tableau*, O'Reilly, 2018), I do always say that if you must use them, stick to five slices or fewer including Other. This chapter demonstrates how to dynamically display the Top *N* based on the number of dimension members the end user wants to show and how to group everything else into a segment for Other.

How to Group the Top N and Everything Else with Parameters and Sets

The first step to showing the dimension members of the Top *N* and grouping everything else into one line is to create a parameter for the Top *N*. This will eventually allow the end user to choose how many individual dimension members to show (i.e., Top 5, Top 6, Top 7).

The allowable options are flexible and up to you, but here's one example in which I allow options between 5 and 20 in multiples of 5:

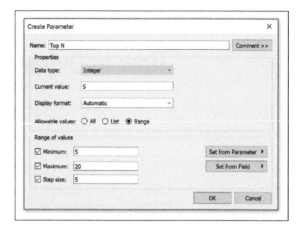

Next, create a set for the dimension you want to segment. For this tutorial, I will create a set out of the State dimension in the Sample – Superstore dataset. The set will be set up to show the Top N States by Profit measure.

To start a set, right-click the dimension and then, on the menu that opens, hover over create, and then click Set. The Create Set dialog box opens in which you can set the conditions for which states should be included in the set. Because we want to keep the Top N based on a specific measure, navigate to the third tab: Top.

Here you can choose a specific number to keep (the default is 10), but you can also click the drop-down arrow located to the right of the hard-coded value and choose our newly created Top N parameter. Based on our parameter settings, this will keep 5, 10, 15, or 20 states in the set depending on the current selection.

To finalize the set, we also want to change the measure that the set is based on to Profit.

Here's how the set looks at this point:

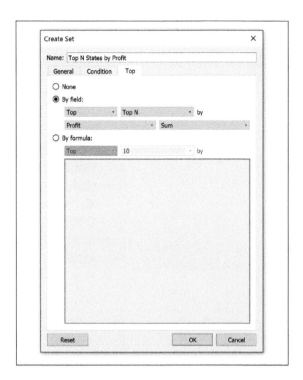

The next step is to create a calculated field that will do the desired grouping. Here is the formula:

```
IF [Top N States by Profit]
THEN [State]
ELSE "Other"
END
```

Now we're ready to use this calculated dimension in a chart. Let's place this new calculation on the Rows Shelf, the Profit measure on the Columns Shelf, and sort the bars by Profit. Also, be sure to show the parameter control by right-clicking the Top N parameter and then choosing Show Parameter Control. This allows you and the end user to choose between the allowable choices of 5, 10, 15, and 20.

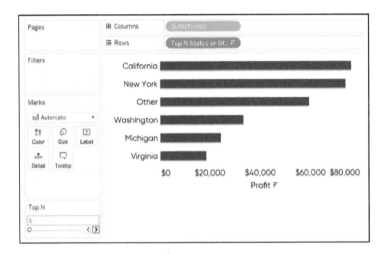

As you can see, when the default parameter value of 5 is selected, there are six bars; one for each of the top five states by profit and one for everything else. If you always want the Other bar on the bottom, place the Top N States by Profit set we created earlier as the first field on the Rows Shelf. This will break up the sets of bars depending on whether they're in the Top N:

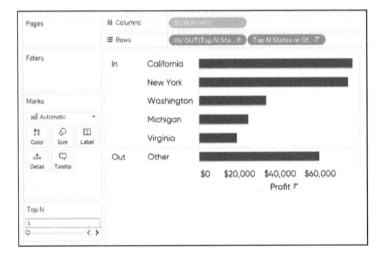

You can always hide the first header for In versus Out by right-clicking it and deselecting Show Header. Now, the end user can choose to group the top 5, 10, 15, or 20 and everything else into its own segment called Other. Here's how the final view looks after I clean up the formatting and change the Top N parameter to 10:

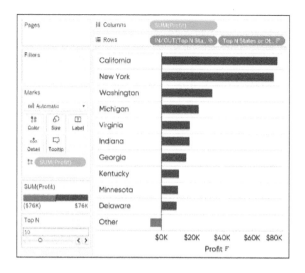

CHAPTER 60

3 Ways to Use the Describe Feature

The *Describe* feature in Tableau is one that somehow took me years to stumble upon but is now one of my favorites. If you've heard of this feature, you've most likely used it to describe an individual dimension, measure, set, or parameter. Right-clicking a field and then choosing Describe gives you information such as the data type and which table it comes from, but there are several other ways to use the Describe feature in Tableau to help you speed up your learning and analyses.

This chapter shows you:

- How to get the most out of the aforementioned usage
- A method for reverse engineering somebody else's (or your own) Tableau workbooks
- A way to view underlying formulas and parameter allowable values.

These tips have several benefits including saving processing time of your VizQL queries, speeding up your learning or reminding you of your previous approaches, and making your calculated field writing more efficient.

This chapter uses the *Super Sample Superstore* dashboard to show you three different applications of the Describe feature in Tableau.

Using Describe to Save Processing Time

As mentioned, if you're familiar with this feature, you've likely right-clicked an individual field and chosen Describe to learn more about the field's properties.

Here's what is shown when doing this for the Segment dimension in the Sample – Superstore dataset:

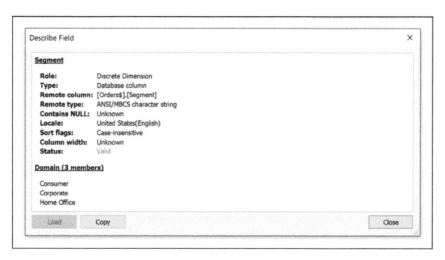

My favorite part of the description when used with a dimension is that Tableau provides a preview of the first 20 dimension members. Before I knew about this feature, I would drag and drop the field onto the view to get an idea of what dimension members were in the field.

Unfortunately, this meant that Tableau had to query the data source to visualize the answer, which could take quite some time depending on the size of my dataset. The Describe feature is much more efficient and gives me an idea of what dimension members are in the field without requiring me to create a new view and wait for processing.

A Method of Reverse-Engineering

Sometimes, I feel like the younger me is smarter than the current me. I don't know whether to laugh or cry when I do a Google search, only to find the answer to my own question in an old blog post I wrote. Other times, I'll go back to old workbooks to remind myself how I executed a certain tactic. I'm also a big proponent of Tableau Public and am always encouraging people to download dashboards they like to "look under the hood" and see how they were created.

Whether you are reverse-engineering your own work or somebody else's, this next tip will help. Even if you've heard of describing fields, you might not know that you can describe entire sheets. To access it, on the menu bar at the top of the Authoring interface, click Worksheet and then, on the menu that opens, choose Describe Sheet. This opens the Sheet Description dialog box.

Here's how a sheet description looks:

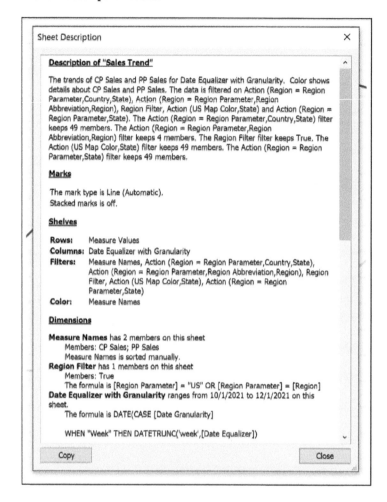

This provides so much information that I actually ran out of room in the screenshot. In addition to the information pictured, you can see all the worksheet's parameters as well as their current values, and even information about the data source being used to generate the worksheet.

You can find most of this information by looking at the pills on the Rows Shelf, Columns Shelf, Filters Shelf, and Marks Cards—but not all of it. For example, you are now able to see whether the marks are stacked and, in this example, view the underlying calculation for the Date Equalizer with Granularity field.

Not only does the Sheet Description dialog box provide a concise total picture of what's happening in the worksheet, but you can use the Copy button located in the lower-left corner to copy and paste the text into a Notepad or Word document. You can then use the code to re-create formulas as calculated fields that you might not otherwise be familiar with.

Calculated Field Writing

Speaking of writing calculated fields, the Describe feature also has "an app(lication) for that." (See what I did there?) For example, instead of getting the formula for the Map KPI calculated field, or just taking a closer look, we can right-click the field and choose Edit. This opens a calculated field dialog box that looks like this:

Clicking anything colored blue, purple, or orange gives you additional information in the *fly-out* box on the right side. Here's what happens when I click `[Parameters]`. `[Map KPI]`:

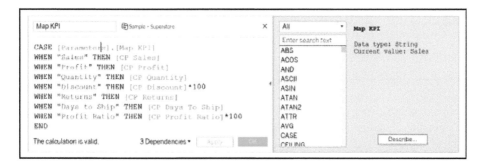

When clicking anything that's purple (parameters) or anything that's orange (fields), you will see the option to describe them in the fly-out box. For parameters, I like to use this Describe button to see the allowable values, which in this case, helps me write the CASE logic:

For calculated fields, I like to use the Describe button to copy and paste underlying code. While you can *view* the underlying code on the right by just clicking an orange calculated field, using the Describe button opens more information, including text that you can copy.

In both of these cases, I'm able to write new calculated fields much faster because I don't need to close the calculated field window I am working in to reopen a parameter to see its allowable values, or reopen a different calculated field to take advantage of code that was already created!

CHAPTER 61

3 Ways to Use Tableau in the Flow

I bet my career on Tableau because of its flexibility, but a close second reason is Tableau's authoring experience. One of my favorite aspects of Tableau's authoring experience is the ability to update my analyses *in the flow*, without having to interrupt my line of thinking or take redundant steps to answer new questions. This provides the benefits of rapid iteration and reducing the risk of distracting stakeholders in my audience.

This chapter shares three of my favorite applications of using Tableau in the flow. I show you a clever way to reverse-engineer table calculations, how to update calculations on the fly, and how to save advanced calculations for future use.

How to Reverse-Engineer Table Calculations

To show you how to reverse engineer a table calculation, I have set up the following chart with one of my favorite applications of this feature. On the first row, we are looking at sales by month; on the second row, we are using a quick table calculation to compute the month-over-month difference in sales.

With this view, the first row is a trend, and in the second row, Tableau is doing the math for us to visualize the month-over-month change:

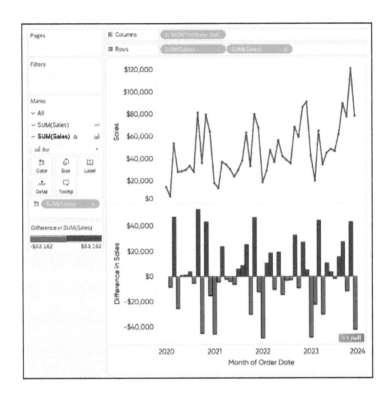

Table calculations are identified with a delta (Δ) symbol on the pill they are being used on, which is why we see this distinction on the second Sales pill on the Rows Shelf. Tableau comes with 11 different *quick table calculations*, including the currently used Difference, that allow you to do some fairly advanced formulas without needing to know how to write the syntax.

It would be great if we could see the syntax so we can reverse-engineer and learn from the computation—which is the first use of Tableau in the flow that I'm going to share!

Double-clicking a pill with a delta symbol displays the underlying formula. Here's how my view looks after double-clicking the second Sales pill on the Rows Shelf:

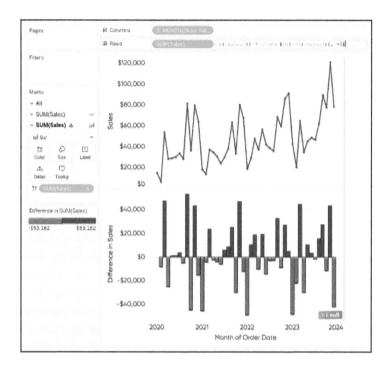

You will see different functions and operators in this formula depending on the quick table calculation being used. If you already know the functions, you can immediately reverse-engineer the formula to determine what's happening on the second row. If the functions are new to you, that's OK; create a calculated field and simply copy and paste the formula into the dialog box:

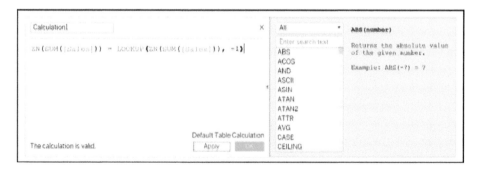

Any time you see something blue in a Tableau calculated field dialog box, you can click it to view a definition in the fly-out box on the right and to see how you should use it in a formula. For example, here's how the definition looks if I click LOOKUP.

After reviewing the formula and the function definitions, I can see that the quick table calculation is taking the value in each cell minus the value at an offset of −1. Because the default addressing of a table calculation is across (from left to right), the offset of −1 means that the value will be from the −1 *column*. From here I can either tweak the formula if needed and/or give it a name and save it for future use.

How to Update Calculations on the Fly

Not only can you see the underlying formula of a table calculation in the flow of your analysis, but you can also *update* a formula. For example, if I were to double-click the pill with a delta symbol, change the offset from −1 to −12, and then click Enter, this would be the result:

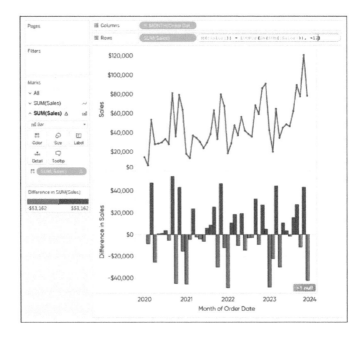

We would now have a year-over-year difference (current month's value minus the value 12 months ago) instead of a month-over-month difference. To get the color to work properly, I had to make this same change to the table calculation on the Color Marks Card. Note the first 12 months are blank because we don't have the historical data to do the year-over-year calculation:

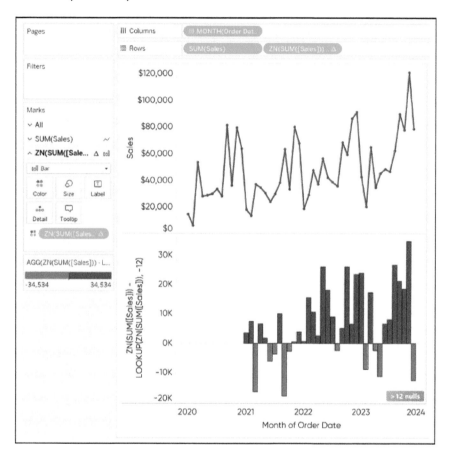

This functionality also works for non-table calculations. This is one of my favorite uses of Tableau in the flow because it helps me to avoid distractions with my end users. For example, if I were to present the preceding chart in a meeting and one of the stakeholders says, "This is a great chart, but I'm much more interested in profit ratio than sales," I can almost instantly update the view by double-clicking the first Sales pill on the Rows Shelf and adding SUM([Profit]) as the numerator:

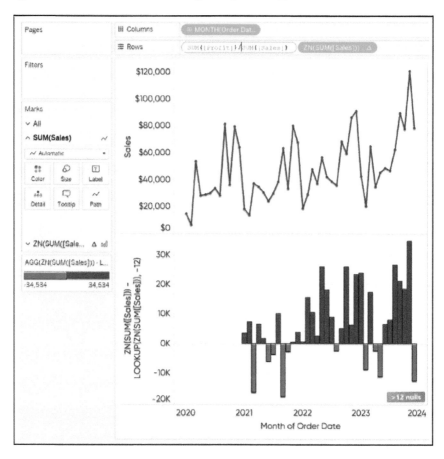

After I click Enter, we have a completely different KPI trend on the first row:

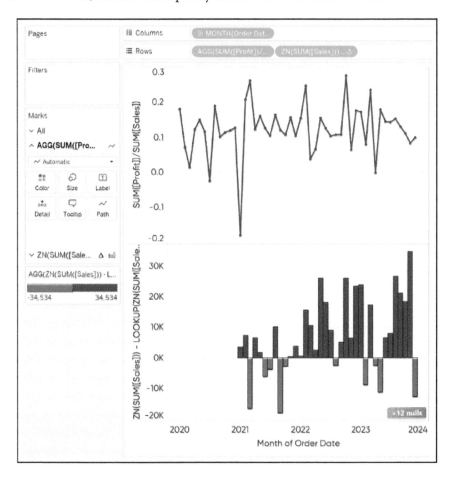

The reason I like this so much is that it removes excuses to slow us down. No, we're not going to "go back and update the data source." No, I don't need to "figure out how to add that KPI to the visualization." No, we don't need to "schedule a series of follow-up meetings to find an insight." We asked a new question in the flow of our analysis and were able to answer it on the fly in just a couple of clicks!

How to Save Updated Calculations for Future Use

I showed you earlier how you can create a new calculated field and copy and paste the code from a table calculation, but there is an easier way to save a calculation for future use. In both the case of a newly created table calculation or an updated calculated field (as we did with profit ratio), you can drag the pill into the Measures area of the Data pane to save it for future use. Here's how the view looks after dragging the second pill on the Rows Shelf to the Measures area of the Data pane and giving the measure a name:

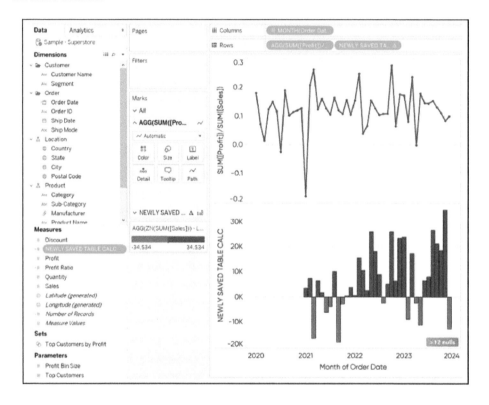

I can now use this table calculation on a different view without having to repeat the steps of adding a table calculation to a raw measure and updating the offset. This is a calculated field now, so I can also update it at any time by right-clicking the field and then choosing Edit.

All three of these approaches improve the authoring experience by allowing you to stay in the flow of your analysis. You can learn more about individual fields, update calculated fields, and even save calculated fields for future use, all without leaving the view you are analyzing!

CHAPTER 62

How to Make a Signature Line with Data Status Alert

The practice of data visualization is very much a psychological exercise. If you can get in the head of your audience and understand their needs and what will resonate with them, you will maximize the chance of your visualization causing action. Some audiences rely on establishing trust with the data and/or the analyst themselves before taking action.

For this reason, I often like to close my dashboards with a *signature line* that usually includes:

- The name of the author and how to contact them
- A list of data sources the dashboard is created with
- A notification that informs the user whether the data is up-to-date

This chapter shares an example of a signature line on a Tableau dashboard and shows you how to create a data status alert so that your stakeholders will always know whether the data source is current.

How to Automatically Detect Whether Data Is Up-to-Date

To introduce the concept of a signature line on a dashboard, take a look at the bottom of the following *Super Sample Superstore* dashboard (*https://oreil.ly/6Unk0*). Only the Explanatory dashboard is pictured, but you can find a similar signature line at the bottom of all three dashboards in the workbook:

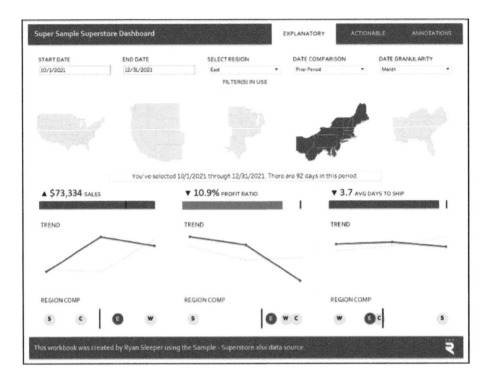

Note I've included the name of the author and the data source with which the dashboard was built. This is a nice touch for amiable and analytical audiences who will want to build rapport with the author and dig deeper into the data, respectively. Something magical happens when these audiences know the dashboard is tied to a real person and was created with data that they can validate.

On internal corporate dashboards, I usually include a way to contact the author (e.g., email address, phone extension) in case there are questions about the data or dashboard. Another nice touch for the signature line is a confirmation that the data is up to date or alerts them when the dataset is no longer current.

To illustrate how to create this alert, I'm going to start a new file with the Sample – Superstore dataset. Let's pretend it's November 7, so for the purposes of illustration, I'm going to filter the sample dataset to include dates through *today*:

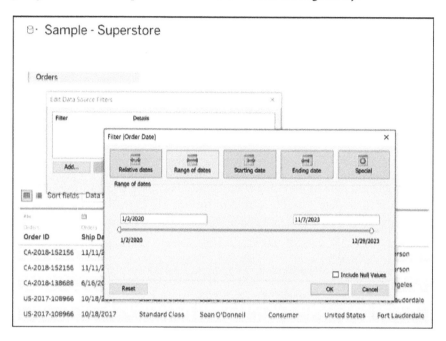

After I'm in the Authoring interface, I'll make the signature line including a data status alert. The data status notification is built by using a simple calculated field:

```
IF DATE(Today()) = {MAX([Order Date])} THEN "Ahh yeah… dataset is up to date."
ELSE "Oopsa - daisy… dataset is out of date!"
END
```

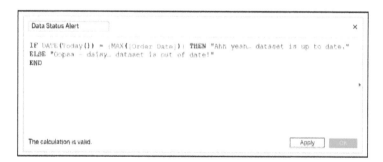

We can also write the formula as follows:

```
IIF(DATE(Today()) = {MAX([Order Date])},"Ahh yeah… dataset is up to date.",
"Oopsa - daisy… dataset is out of date!")
```

Of course, you can replace the message between the quotes with whatever you would like. After you have the calculated field, place it on the Text Marks Card:

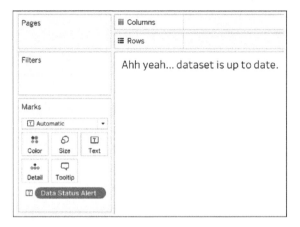

Because the highest date in the dataset matches today's date, the notification confirms that the data is up to date. For illustration purposes, let's change the calculated field to compare the max date to tomorrow's date (making the data source out of date):

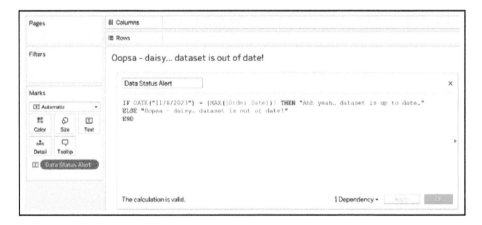

If your data is updated only once per day, meaning that the most current data would actually be yesterday's date, simply add –1 after TODAY() in the first line of the calculated field as follows:

```
IF DATE(Today())-1 = {MAX([Order Date])} THEN "Ahh yeah... dataset is up to date."
ELSE "Oopsa - daisy... dataset is out of date!"
END
```

When you have this formula on the Text Marks Card, you can provide additional information by clicking the Text Marks Card and then clicking the ellipsis to the right of the Text box:

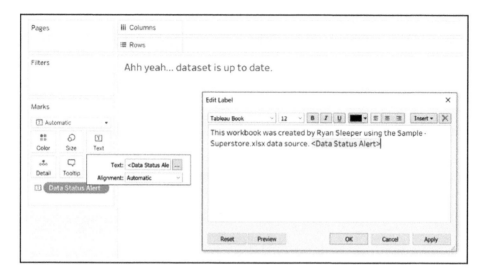

Here is how the final signature line with the data status notification looks after clicking OK:

This workbook was created by Ryan Sleeper using the Sample - Superstore.xlsx data source. Ahh yeah... dataset is up to date.

You can now add this signature line to your dashboard(s) and format it as you like!

CHAPTER 63

How to Improve Your Dashboard Design with White Space

Another of my favorite dashboard elements is less tangible because...well, it's sort of invisible. I'm talking about *white space*, or *negative space*; an important design element that you should consider in the context of your data visualization. Although there is no ink in white space, its use can make or break how your work is perceived and used.

White space helps you as the dashboard author prioritize content, helps your end user focus their attention during their analysis, and adds some professional design polish that will lend itself to better credibility with your audience. This chapter shows you three different tactics for improving the white space in your Tableau dashboards.

What Is White Space in Data Visualization?

Before I share my three tactics for implementing white space in your Tableau dashboards, look at this example to see the difference that this negative space can make. In this first image, I've used mostly default spacing, padding, and layout.

Note this could be even worse because I left the nicely spaced navigation buttons and centered text in the date caption. Tableau also tries to help you with this very design aspect we're discussing because, by default, it adds four pixels of padding around the vertical layout containers that are holding the individual sheets related to each KPI.

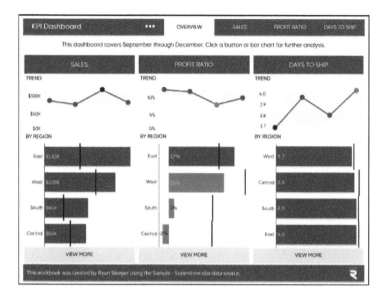

Still, you can see this isn't ideal. Objects are too close together; it's not clear whether you should be reading left to right or from top to bottom; it's not easy to see there are buttons to drill into deeper analyses.

Generally, this looks like a big block of stuff, and I'm not sure whether I trust the dashboard's author or its underlying data. Here is the same view after implementing the three tactics I'm about to share:

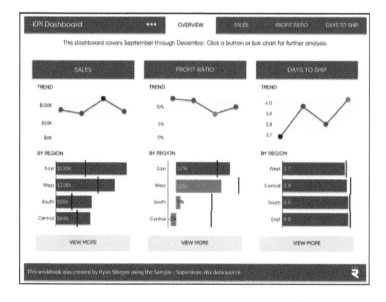

This dashboard has the exact same elements, but the design is much nicer and easier to consume. The only difference between the two is white space.

Sketch Out the Dashboard Dimensions in Advance

In my opinion, the best way to ensure that your dashboards have a nice layout that incorporates white space, is to sketch out the dashboard dimensions in advance. For example, if I know that the previous example dashboard was 1,024 pixels wide overall, would contain three layout containers across, and that I would like each object to have 25 pixels between them, I would do the math to calculate how wide each of those three layout containers should be. Let's look at that layout calculation in a more methodical manner:

[25 px white space] [Container 1] [25 px white space] [Container 2] [25 px white space] [Container 3] [25 px white space]

So, I know there is a total of 100 pixels of horizontal white space. I would subtract that amount from the overall dashboard width, which leaves me with 924 pixels for the three containers. I would then divide that number by three, which tells me each layout container should be 308 pixels wide. Here's a sketch of how all the dashboard elements might look:

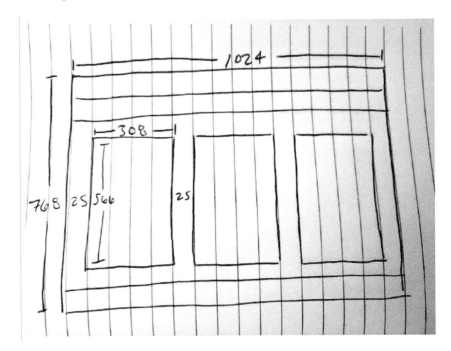

And I really do something like this; either in a notebook or on a white board. After I have the sketch with the dashboard dimensions, it's very easy to add objects to the view and change their precise sizing using the Layout pane:

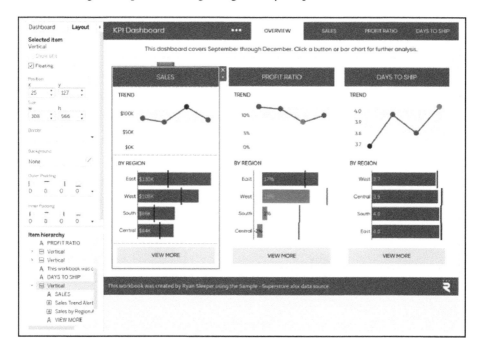

Take Advantage of the Padding Options on the Layout Pane

I mentioned that Tableau automatically adds 4 pixels of padding around dashboard objects, but you can adjust this to your liking by clicking any dashboard object and then, in the upper-left corner of the Dashboard interface, click the Layout tab and update the Outer Padding and/or Inner Padding:

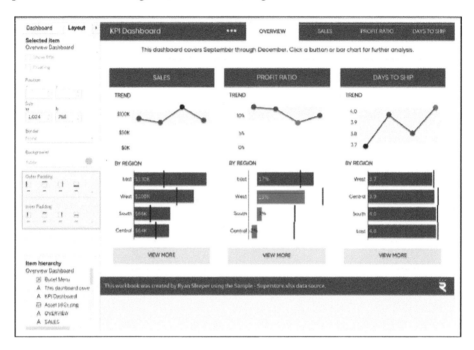

I began sketching out dashboard dimensions with the arrival of Tableau 8.0, which introduced *freeform dashboard design*, or the ability to use both tiled and floating objects. At the time, this was the only way to completely customize the white space throughout a dashboard, but that changed with the introduction of object padding in Tableau 10.4.

Many of the same formatting options that I would need to tediously and manually create before are now just a couple of clicks away on the Layout pane. Before this feature was added, I was firmly on Team Floating, but now I'm closer to Team Hybrid. I will often make floating layout containers, but the objects within those containers are tiled. I then add padding to create the desired white space.

Use a Blank Dashboard Object

If you are still on Team Tiled, I'll close with the white space object that started it all: Blank. You can add a Blank object—which does nothing but add white space between tiled objects—by dragging the Blank object from the Dashboard pane onto a dashboard. See the Sales container in the following screenshot to see what this looks like:

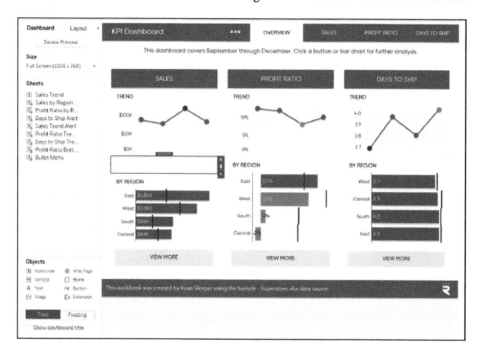

The one drawback with this approach is that you cannot precisely control the size of the Blank object. You can resize it, but you don't have the same pixel-level precision that you get with floating objects. This is another reason I was on Team Floating and now Team Hybrid.

Whichever approach you take, considering the white space of your Tableau dashboards will help declutter your work and communicate your insights in a professional way.

CHAPTER 64

How to Display Varying Decimal Places

Tableau inspiration and problem solving can come from anywhere, and this tactic originated from a question from a Twitter connection! The user was trying to display four decimal places when the measure value on the view was less than one, but only two decimal places when the measure value was greater than one. This was a brain-teaser because, by default, Tableau limits you to one number format per measure.

I loved the concept because I'm a big believer in maximizing the data-ink ratio, and the extra decimal places could be considered redundant data ink. My first instinct was to dynamically format the numbers with parameters, but this technique works only for controlling the prefix and suffix of each different number type. Instead, this chapter shows you two alternative approaches that allow you to control the prefix, suffix, and/or the number of decimal places when you are trying to display two (or even three!) number formats for the same measure in Tableau.

 Related: *Practical Tableau*, Chapter 65, "How to Dynamically Format Numbers" (O'Reilly, 2018)

Using Tableau's ROUND and IIF Functions to Display Varying Decimal Places

If your only objective is to round decimal places based on a measure's value, you can use Tableau's ROUND and IIF functions in a calculated field.

To illustrate the dynamic formatting, I've put together this dummy dataset with "10 things." Note the measure values include varying levels of decimal places. You can view and download this dataset from Playfair Data (*https:// oreil.ly/TTJ52*):

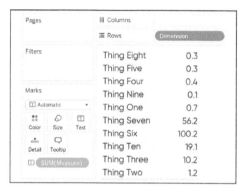

	A	B
1	Dimension	Measure
2	Thing One	0.67391029
3	Thing Two	1.208381
4	Thing Three	10.23408
5	Thing Four	0.4230283
6	Thing Five	0.334810235
7	Thing Six	100.2134098
8	Thing Seven	56.15093248
9	Thing Eight	0.29804315
10	Thing Nine	0.093246851
11	Thing Ten	19.12350981

If I were to make a table with these numbers in Tableau, the measure values would have Automatic formatting applied and every number would display one decimal place:

I could either change the default formatting of the measure or change the format of the measure for this specific view to change the number of decimal places, but in both cases, all the numbers would share the same number of decimal places. If we are trying to display four decimal places for values that are less than one, and two decimal places for numbers that are greater than one, we can use the following calculated field:

```
IIF([Measure]<1,ROUND([Measure],4),ROUND([Measure],2))
```

This formula looks at the first statement, [Measure] < 1, and when that statement is true, Tableau rounds the original measure to four decimal places. If that first statement is false, it rounds the measure to two decimal places. Now, replace the original measure from the dataset with the newly created calculated field.

By default, you will still see Automatic formatting and the measure values will be rounded to one decimal place. However, now when you change the number formatting to Number (Standard), you will see the dynamic formatting appear. To find this formatting, right-click the measure that is currently on the Text Marks Card, and then, on the menu that opens, choose Format, and then change the format for the numbers on the *Pane* (as opposed to Axis). Values below one are rounded to four decimal places, and values above one are rounded to just two:

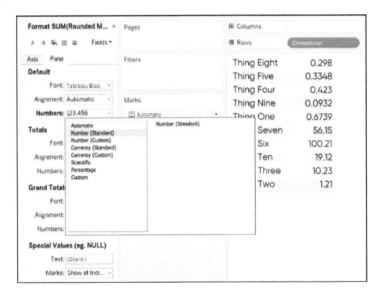

The benefit of this approach is that it works whether you're formatting positive or negative values and elegantly solves the question at hand regarding decimal places. Read on for a solution that addresses broader applications.

How to Display Two (or Three) Number Formats for the Same Measure in Tableau

This next approach is more involved but unlocks more flexibility, including the ability to show a third variation as well as different number formats. I learned this trick from Tableau Zen Master Hall of Famer Andy Kriebel in his post "One Metric; Two Number Formats" (*https://oreil.ly/5sbwF*). Andy used the technique to display different number types (e.g., percentages, currencies) when you are allowing users to select measures with parameters, but this also works for displaying different numbers of decimal places for the same measure.

First, create a calculated field that multiplies whatever you want to have a unique formatting by negative one.

For this scenario, let's assume that two decimal places will be our default, but we want to display four decimal places for values less than one. Therefore, the values less than one are the ones we multiply by negative one. Here's the formula:

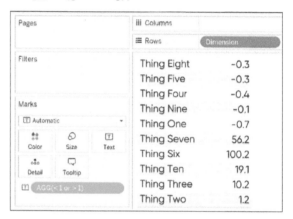

```
IF SUM([Measure]) < 1 THEN -SUM([Measure])
ELSE SUM([Measure])
END
```

We can also write this formula as follows:

```
IIF(SUM([Measure])<1,-SUM([Measure]),SUM([Measure]))
```

Next, replace the original measure with the newly created calculated field that converts your exception to negative:

Now that we have a mix of negative and positive values, we can apply a unique format to both types of values. To do this, right click the measure on the view (currently on the Text Marks Card in my example), on the menu that opens,

choose Format, and then update the format for the numbers on the Pane. When you choose Custom, any format you type before a semicolon will be applied to positive numbers, and any format you type after a semicolon will be applied to negative numbers:

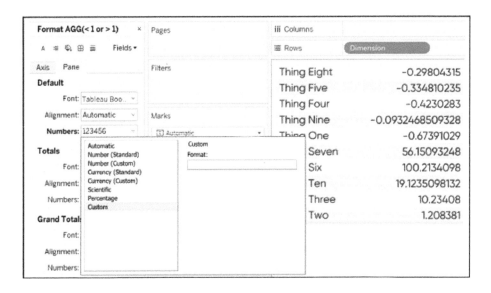

If I type .00 before a semicolon in the custom text box, that number formatting will be applied to positive numbers. In my example, the positive numbers are the values above one. If I type .0000 after a semicolon, that format will be applied to negative numbers, or the values below one in my example:

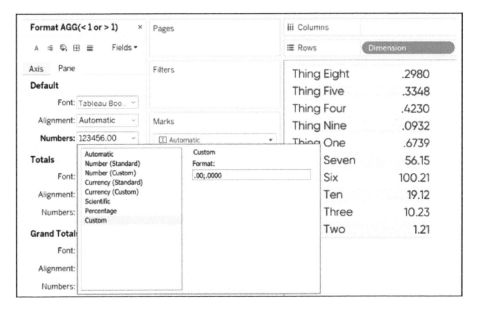

Take a closer look at the table in the last image; values less than one display four decimal places, and values greater than one display just two! Again, we're using this as an alternative solution to rounding decimal places, but you can create any second number formatting for your exception (the measure being multiplied by negative one). This could be different decimal places, a dollar sign, a percent sign, and so on.

But wait—there's more!

You can unlock a third number format by typing a second semicolon in this text box for custom number formatting. This is the format that will be displayed for values that equal zero. Suppose that we don't want to focus on the dimension members in our view that have already reached their goal for the year. Instead, I will add a star alt-code character (★) to communicate that we no longer need to worry about those items.

First, I need to add a line of logic to my calculated field that converted some values to negative to also convert some of our values to zero. Let's say our goal for the year is 50, so any dimension member that has reached that threshold will be converted to zero. Here's the formula:

```
IF SUM([Measure]) < 1 THEN -SUM([Measure])
ELSEIF SUM([Measure]) > 50 THEN 0
ELSE SUM([Measure])
END
```

In addition to displaying four decimal places for values less than one, and two decimal places for values greater than one, this formula now converts the values that are past our goal of 50 to zeros. Now if I go type a second semicolon into the custom formatting for our calculated measure, whatever number format I type next will be applied to the values that are above goal.

Here's how the view looks when typing a star alt-code character as the third number formatting:

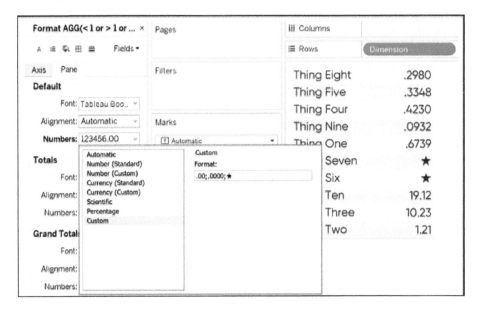

If you would still like to display the actual value for any of these dimension members on the table, you can simply add the original measure to the Tooltip Marks Card. In other words, you can display a star to let the end user know that they don't need to worry about that item, but if they want to be reminded of the actual performance, they'll see it when they hover over the marks.

The only caveat with this trick is that you can't have negative values or zeros, because this new custom format would be applied to those actual negatives and zeros.

CHAPTER 65

How to Dynamically Change Number Units Between K/M/B

This chapter shows you how to automatically change number formatting in Tableau between no units for values less than one thousand, K for thousands, M for millions, and B for billions. Number display units are a great way to clean up visualization labels or to save real estate in a text table if you are looking to maximize the data-ink ratio described in Chapter 12.

The issue is that number formatting is all or nothing per measure, so if you set the formatting to display units of K for thousands but then have a value of one hundred, Tableau will display 0K. Not anymore—there's a hack for that!

How to Automatically Change Display Units for Thousands, Millions, and Billions in Tableau

Because formatting is per measure in Tableau, I've previously covered how to dynamically format measures, even when your field requires more than two number formats (i.e., integers, currency, and percentages).

 Related: *Practical Tableau*, Chapter 65, "How to Dynamically Format Numbers" (O'Reilly, 2018)

The trick to dynamically changing display units between none, K, M, and B follows a similar approach, but instead of a calculation for the number prefix and suffix, you need a calculated field that normalizes each measure and a second calculated field for the suffix (which becomes the display units).

If you want to dynamically change display units for the Sales measure in the Sample – Superstore dataset, the first formula would be as follows:

```
IF SUM([Sales]) >= 1000000000 THEN SUM([Sales]) / 1000000000
ELSEIF SUM([Sales]) >= 1000000 THEN SUM([Sales]) / 1000000
ELSEIF SUM([Sales]) >= 1000 THEN SUM([Sales]) / 1000
ELSE SUM([Sales])
END
```

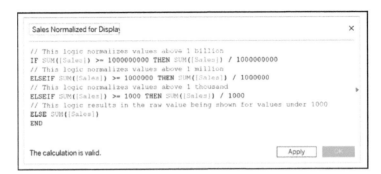

For this example, let's also set the default formatting of this calculated field so that it always has a dollar sign and two decimal places. To modify the default formatting of a calculated field, in the Measures area of the Data pane, right-click it, hover over Default Properties, and click Number Format. This opens the Default Number Format dialog box.

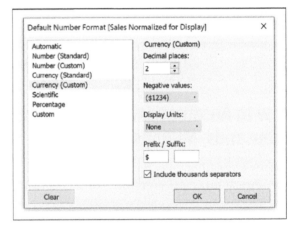

The prefix will always be a dollar sign but leave the suffix blank because that is what we will dynamically control with the next step.

The second calculated field displays the appropriate format based on the measure value. Here's the formula to show B for billions, M for millions, and K for thousands:

```
IF SUM([Sales]) >= 1000000000 THEN "B"
ELSEIF SUM([Sales]) >= 1000000 THEN "M"
ELSEIF SUM([Sales]) >= 1000 THEN "K"
ELSE "
END
```

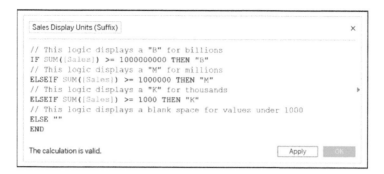

```
Sales Display Units (Suffix)                                    ×

// This logic displays a "B" for billions
IF SUM([Sales]) >= 1000000000 THEN "B"
// This logic displays a "M" for millions
ELSEIF SUM([Sales]) >= 1000000 THEN "M"
// This logic displays a "K" for thousands              ▶
ELSEIF SUM([Sales]) >= 1000 THEN "K"
// This logic displays a blank space for values under 1000
ELSE ""
END

The calculation is valid.                    Apply    OK
```

To use this in a text table, place the normalization and suffix calculated fields on the Text Marks Card together. You can format the text so that the two results from the calculations are on the same row, making them appear seamless. Of course, you can also format the text for style such as font, size, and color:

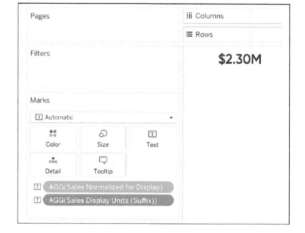

Even though the Display Units (Suffix) calculated field always results in text, you will find it in the Measures area of the Data pane because the result is dynamic.

Because we haven't specified any level of detail more granular than the entire Sample – Superstore dataset, we are looking at the total Sales amount for the file, which is $2,297,201. Because this number is greater than one million, our normalization calculated field divided it by one million and our suffix calculated field displayed an *M*.

If I add the Region dimension to the view, which has dimension members with values between one thousand and one million, the calculations work to properly normalize the values into thousands and display a *K*:

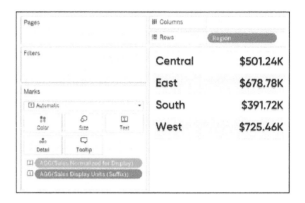

Note that the normalization and suffix calculated fields compute the results at the viz level of detail, so be aware that you can get a mix of results, which can be confusing for your audience. Here's an example when using Ship Mode as the dimensional breakdown:

Although I like display units because they reduce redundant data ink, and I particularly like the tactic being shared to dynamically change display units for its convenience, there are a couple of pitfalls to beware of. Because the normalization calculation happens at the viz level of detail, you might end up with some dimension members being divided by different numbers, which can be misleading to your audience. This tactic works best if all dimension members will have the same tier of results (i.e., billions, millions, or thousands).

The misleading effect is particularly problematic for visualizations because every number, regardless of how it was normalized, will share the same axis scale. For example, if one bar has been divided by one million while another bar was divided by one thousand, the best-performing dimension member will appear to have performed the worst! That being said, this is still an effective tactic for dynamically changing *labels* on visualizations (as opposed to marks).

For the last example, let's break the Normalized Sales measure down by the Product Name dimension:

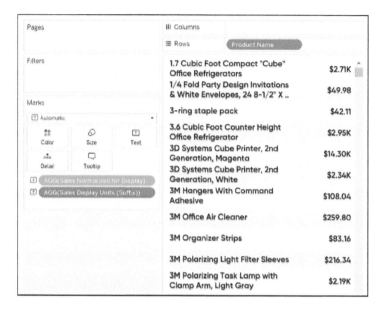

As you can see, whenever a product has a thousand dollars of sales or more, our calculations normalized the result into thousands with a K. The values under one thousand dollars in sales get no display units and the raw result in currency formatting is displayed.

The Sample – Superstore dataset does not have an example of Sales greater than one billion, but our calculations would work the same way as you've seen here with millions and thousands.

This approach also theoretically works with trillions. If your company is working with numbers in the trillions, you should get in touch with Playfair Data!

CHAPTER 66

3 More Innovative Ways to Use Dates: Tip 1

How to Compare the Last Two Full Days, Weeks, Months

This chapter aims to help you harness dates in Tableau to create powerful comparisons in your dashboards. You will learn how to isolate the last two full reporting periods—whether they be days, weeks, months, quarters, or years—so that you can compare the last complete date part to the date part preceding it (e.g., last week compared to the week before). You can use the calculations shared in this chapter as a foundation to create period-over-period percent or index changes, filter your dashboards to only the most recent dates, and normalize the dates so they overlap on the same axis.

Even though there is almost always more than one way to do the same thing in Tableau, I've attempted to provide an easy-to-execute solution that also processes efficiently. As such, this approach allows you to compare the last two complete date parts without the use of level of detail calculations or table calculations. I owe a big thank you to Playfair Data partner consultant, Rody Zakovich; he collaborated with me on this chapter to make the calculations even more elegant than my original idea.

How to Isolate the Last Two Complete Date Periods

This technique works regardless of what date part you are using (day, week, month, quarter, or year), assuming that your data source is updated at least daily. The foundational calculated field that you need computes whether the date range is within the last full period or two full periods ago. You can write this formula with CASE WHEN or IF THEN logic, but the CASE WHEN code is slightly more concise. To illustrate, let's use the date part 'week' in the Sample – Superstore dataset.

 As we progress, keep in mind that we're pretending to write this on 12/22/2021.

Here's the formula:

```
CASE [Order Date] < DATETRUNC('week', TODAY())
WHEN [Order Date] >= DATEADD('week', -1, DATETRUNC('week', TODAY())) THEN "Full
Current"
WHEN [Order Date] >= DATEADD('week', -2, DATETRUNC('week', TODAY())) THEN "Full
Prior"
END
```

Full Current / Full Prior

```
CASE [Order Date] < DATETRUNC('week', TODAY())
WHEN [Order Date] >= DATEADD('week', -1, DATETRUNC('week', TODAY())) THEN "Full Current"
WHEN [Order Date] >= DATEADD('week', -2, DATETRUNC('week', TODAY())) THEN "Full Prior"
END
```

The calculation is valid.   Apply   OK

The first line in this logic limits the date range to dates that are less than today's date, truncated at the week date part, which eliminates partial weeks. If today is 12/22/2021 and the week starts on Monday, all dates would need to be less than 12/20/2021.

 By default, weeks in Tableau begin on Sunday. To change the start day for your weeks, on the menu bar at the top of the window, click Data, hover over the name of the data source, and then choose Date Properties.

The next line in the calculation specifies that the date must be greater than or equal to the start of the last full week. The third line says the date must be greater than the start of the second last full week from today. You might notice that there is some overlap in dates between the second and third line in this statement (today is both greater than one full week ago and two full weeks ago), but each date can be classified only once. This means that after a date is classified as "Full Current", it cannot be reclassified as "Full Prior".

If you prefer to think in terms of IF THEN logic, here's how the formula would look:

```
IF [Order Date] < DATETRUNC('week', TODAY()) THEN
(IF [Order Date] >= DATEADD('week', -1, DATETRUNC('week', TODAY())) THEN "Full
Current"
ELSEIF [Order Date] >= DATEADD('week', -2, DATETRUNC('week', TODAY())) THEN "Full
Prior" END)
END
```

Whether I'm using the CASE WHEN or IF THEN version, if today's date is 12/22/2021, the last full week should start on 12/13/2021, and two full weeks ago should start on

12/6/2021. Let's see if it works. Make a *sales by day* line graph filtered to December 2021 and put the newly created Full Current / Full Prior calculation on the Color Marks Card:

As you can see, the last full week is colored blue and starts on 12/13/2021; the second full week ago is colored red and starts on 12/6/2021; everything else is classified as Null and is colored gray. We have just isolated the last two complete reporting periods! This calculation works if you replace the 'week' date part with 'day', 'month', 'quarter', or 'year'.

Other Ideas for Comparing the Last Two Complete Date Periods in Tableau

There are several tactics we can implement from this point, including filtering the view to only the last two complete periods, creating period-over-period percent change calculations, normalizing the comparison ranges so that they overlap on the same axis, and parameterizing the date part to change between days/weeks/months/quarters/years on the fly.

Using the Full Current/Full Prior dimension as a filter

In the previous screenshot, you can see that the dates that are not classified as Full Current or Full Prior are called Null and colored gray. To remove these dates from the view to focus on the relevant dates only, add the newly created calculated dimension to the Filters Shelf and then clear the Null checkbox:

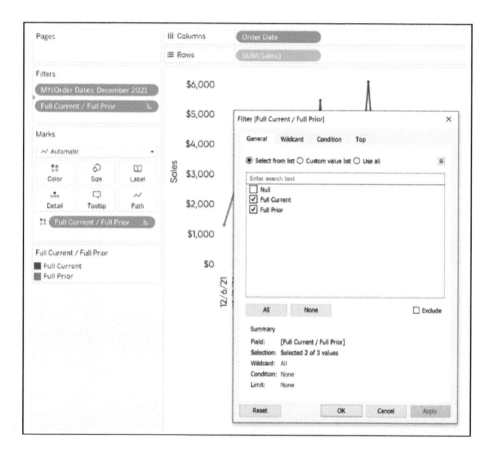

By the way, if you would prefer your null dates to have a better name, you can add an ELSE statement at the bottom of your CASE WHEN or IF THEN calculated field. For example, I might prefer to call my null dates "Not in Range" by adding one more line to my calculated field:

```
CASE [Order Date] < DATETRUNC('week', TODAY())
WHEN [Order Date] >= DATEADD('week', -1, DATETRUNC('week', TODAY())) THEN "Full
Current"
WHEN [Order Date] >= DATEADD('week', -2, DATETRUNC('week', TODAY())) THEN "Full
Prior"
ELSE "Not in Range"
END
```

You can similarly limit the dates on the view by using a relative date filter in Tableau, but the difference is that the version I'm sharing filters out partial weeks.

Compute percent or index changes between the last two full periods

Isolated numbers do not provide much value, and it is almost always better to provide context in the form of a comparison. I'm such a big believer in this that my first go-to dashboard element is the current versus comparison index callout described in Chapter 39. It is very easy to create this powerful component now that we have the foundational Full Current / Full Prior calculated field.

Suppose that we want to compare the Sales measure from the current full week to the prior full week. We would need to isolate Sales from Full Current week and Full Prior week; then, compute the percent change or index score.

Here's the formula for Full Current Week Sales:

```
SUM(
IF
[Full Current / Full Prior]
= "Full Current"
THEN [Sales] END)
```

The formula for Full Prior Week Sales is as follows:

```
SUM(
IF
[Full Current / Full Prior]
= "Full Prior"
THEN [Sales] END)
```

And here's the formula to compute the percent change between the last two full weeks:

```
([Full Current Week Sales]
/ [Full Prior Week Sales])
- 1
```

Here's how the *current versus comparison callout* could look after creating these three calculated fields:

Equalize the comparisons so they line up on the same axis

If you're comparing the last full week to the full week two weeks ago on a line graph, it would be easier to compare day over day (i.e., Tuesday versus Tuesday) if the two lines overlapped on the same axis.

To accomplish this, instead of using the default date field in your dataset on the Columns Shelf, replace it with a Date Equalizer calculated field that moves the dates from the comparison period forward on the x-axis. For example, if the date part is week, the formula would be this:

```
IF [Full Current / Full Prior] = "Full Prior" THEN [Order Date] + 7
ELSE [Order Date]
END
```

You can also write this formula as follows:

```
IIF([Full Current / Full Prior] = "Full Prior",[Order Date] + 7,[Order Date])
```

This formula looks to see whether the date has been classified as Full Prior. If it does, it adds seven days; otherwise it shows the default date. When using a date part of week, the Monday from two full weeks ago (12/6/2021) would be moved up seven days so that it lines up with the Monday one full week ago (12/13/2021).

Here's how my line graph looks when replacing [Order Date] with [Date Equal izer] on the Columns Shelf:

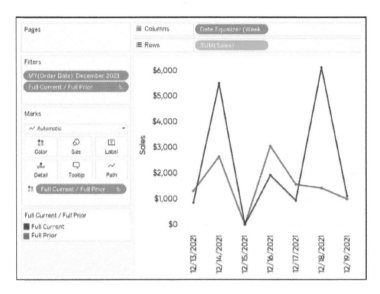

This type of normalization is relatively straightforward when dealing with days, weeks, or years because the ranges contain an equal number of days. Months, quarters, and custom ranges, on the other hand, can have a different number of dates (e.g., February versus January), so you must compute the number of days in the range before adding them to the comparison range to equalize the dates.

For more on normalizing dates and the DATEDIFF function, review how we normalized dates on the same axis in Chapter 53.

Create a parameter to change the comparison date part on the fly

In the examples thus far, I've used the 'week' date part to compare the last full week to the full week preceding it, but I also mentioned the calculation works if you replace week with day, month, quarter, or year. Well, why not allow yourself and your end users to quickly pivot through the different date parts?

This is relatively easy to accomplish by setting up a parameter with a data type of String and a list of allowable values. Each allowable value should represent one of your date parts. The value itself needs to be lowercase and singular, but the Display As value can be title cased or abbreviated:

You would then replace every occurrence of the date part, 'week', in the Full Current / Full Prior calculated field with this newly created parameter, right-click the parameter, and then choose Show Parameter Control. After following these steps, you and your end users will have direct access to choose the date part that is populating the comparison periods.

To take this idea to a Jedi level, see how to use Tableau Parameter Actions to change date parts in Chapter 70.

If you would prefer to automatically compare *partial* date periods, you will find the solution in Chapter 67.

3 More Innovative Ways to Use Dates: Tip 2

How to Compare the Last Two Partial Weeks, Months, Quarters, or Years

This chapter shows you how to automatically isolate the last two *partial* date ranges so that you can compare an equal number of days period over period. In Chapter 66, I show you how to compare the last two *complete* date periods in Tableau, but sometimes you want the comparison to be even more current. For example, you might want to compare this week to last week even if the week is not yet complete instead of comparing the last full week to the full week from two weeks ago.

The benefit to this is timelier analysis, but you can often end up with an apples-to-oranges comparison. For example, if it's Wednesday this week and you try to create a week-over-week analysis, you can end up comparing the three days from this week to all seven days from last week. The tricky part to creating a true partial period-over-period analysis is that you need to calculate the number of days in the current range and then cap the last full date range at that same number.

This chapter shares the formulas needed to automatically create a partial period-over-period analysis so that both periods—whether you're using weeks, months, quarters, or years—contain an equal number of days.

The core of this technique is the same as when comparing the last two complete date parts in that we need to create a calculated field that classifies dates as *current* or *prior*. Partial periods are more nuanced, however, because they can contain any number of days from 1 to 364. You could be comparing anything from Monday this week to Monday last week or January 1 through December 30 this year to January 1 through December 30 last year.

To account for these differences, we must compute the number of days in the current partial date range so that we can ensure the current and comparison ranges contain the same number of days. So, in addition to our Partial Current / Partial Prior calculated field, we need two additional calculated fields: an LOD calculation to automatically compute the latest date in the dataset and a DATEDIFF calculation to determine the number of days in the current partial range.

Here are a couple of notes that will be helpful in case you want to follow along with the Sample – Superstore dataset:

- To create the illustration that we're working with currently and to exclude sample dates that go into the future, I added a data source filter to bring in only dates through Wednesday, December 22, 2021.
- By default, weeks begin on Sunday in Tableau, but I usually think of weeks as starting on Monday. I have adjusted my weeks to start on Monday by clicking Data on the menu bar at the top, hovering over the name of the data source, and then changing the Date Properties.

For this example, suppose that it's Thursday, December 23, and our company updates our data once per day overnight, meaning that our dataset is current through Wednesday, December 22. We would like to create a report that automatically compares the number of days that have passed in the current week (Monday, Tuesday, and Wednesday) to the same three days in the prior week week.

The first calculated field needed is a simple LOD calculation to automatically isolate the highest date in the data. Here's the formula:

```
{MAX([Order Date])}
```

Note this calculation includes a FIXED LOD expression, but because FIXED is the default expression and we are not specifying a dimension to change the LOD expression, we do not need to type **FIXED:** at the beginning of the syntax.

The next calculation computes the difference between this calculated field, or the latest date in the dataset, and the beginning of the current partial date range:

```
DATEDIFF(
'day',DATETRUNC(
'week',[Max Date]),
[Max Date])
```

The first date part of 'day' ensures that the difference between the max date and the start of the range is always computed in days and should not be changed. The second

---

date part—'week' in this case—can be changed to month, quarter, or year depending on what partial ranges you're hoping to compare (partial week over partial week, partial month over partial month, etc.).

Now, we are ready to create the foundational Partial Current / Partial Prior calculated field that drives all the date range classifications. As with the Full Current / Full Prior calculation shared in Chapter 66, you can write this calculation using CASE/WHEN or IF/THEN logic. Although I primarily used CASE/WHEN logic for the example in the previous chapter because it was more concise in that scenario, this time IF/THEN contains the exact same number of characters. I also think it's easier to understand what's happening with IF/THEN logic, so let's use that; the formula thus is as follows:

```
IF [Order Date] <= [Max Date]
AND [Order Date] >= DATETRUNC('week', [Max Date]) THEN "Partial Current"
ELSEIF [Order Date] >= DATEADD('week', -1, DATETRUNC('week', [Max Date]))
AND [Order Date] <= DATEADD('week', -1, DATETRUNC('week', [Max Date])) + [Days in
Range] THEN "Partial Prior"
END
```

This one might look a little bit intimidating, but there are still just two outcomes:

- If the date is less than or equals the latest date in the dataset and the date is greater than or equal to the start of the current partial period, it's classified as Partial Current.

- If the date is greater than or equal to the start of current partial period minus one date part (so the start of the last partial period) and less than or equal to the start of the last partial period plus the number of days in the current partial period, it's classified as Partial Prior. It's the last part of this calculation, + [Days in Range] that caps the prior comparison at the same number of days as the current comparison. Even though there were seven full days last week, we want to compare only the first three in this case.

To illustrate comparing a partial current week to a partial period with the same number of days from the previous week, here's a line graph showing Sales by Order Date in the Sample – Superstore dataset. I've also filtered the view to include only dates that have been classified as either Partial Current or Partial Prior and placed the Partial Current / Partial Prior calculated dimension on the Color Marks Card:

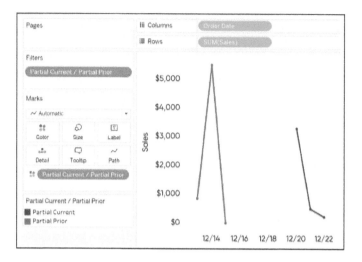

We are left with Monday, 12/20, through Wednesday, 12/22 (the latest date in the data), as our current partial range and Monday, 12/13, through Wednesday, 12/15 as our prior partial range. We just created a partial date range comparison containing an equal number of days that will automatically update as our data refreshes!

To update these formulas to work with months, quarters, or years, simply update (or parameterize) all of the date parts in the Partial Current / Partial Prior calculated field and the *second* date part in the Days in Range calculated field (but leave 'day' as is). Here's another example after updating the calculations to compare partial month over partial month:

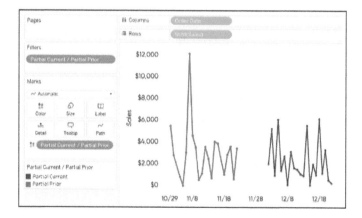

After changing the date part to 'month', the December range spans from December 1 through December 22, and even though the entire month of November has passed in this example, the comparison month is capped 22 days into the month to create an equal number of days. There are some missing dates in the Sample – Superstore dataset, but both ranges span from the beginning of their respective month to the 22nd of the month.

From here, you can do the following:

- Isolate partial current and partial prior measures to create percent change comparisons
- Normalize the partial prior dates so they line up with the partial current dates on a line graph
- Parameterize the date part selection to quickly pivot through different date comparisons

These tactics work the same way as when you're comparing complete date ranges as we demonstrate in Chapter 66.

CHAPTER 68

3 More Innovative Ways to Use Dates: Tip 3

How to Automatically Change Date Granularity Based on Days in Range

I've shared before how to let your end users change the date part of a line graph in Tableau (see Chapter 66 from *Practical Tableau*, O'Reilly 2018). Then this tactic was made even better with Tableau Parameter Actions, which allow the user to change the date part using a button (Chapter 70). These are both useful techniques for quickly making line graphs more and less granular in Tableau.

Well, let's make this tactic better once again by *automatically* changing the date granularity to the most appropriate date part based on the number of days in the selected date range. This chapter shows you how to change the date part of a line graph from day, to week, to month based on whether there are 30 or fewer, 90 or fewer, or more than 90 days on the view, respectively.

This approach requires five short steps:

1. A start date parameter
2. An end date parameter
3. A calculated field to compute the number of days in the selected range
4. A date filter that keeps only the selected dates on the view
5. A calculated field to appropriately truncate the date being used in a line graph

If you are already using parameters to normalize your date ranges on the same axis as is described in Chapter 53, you will have already completed the first three steps.

First, set up your start date and end date parameters. These are simply parameters created with a data type of Date; here's how mine look with a default date range of 5/1/2023 to 5/7/2023:

1. Start Date parameter:

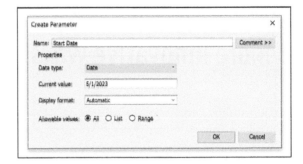

2. End Date parameter:

Next, create a calculated field that computes the number of days in this selected range. The formula is simply ([End Date] - [Start Date])+1. The +1 ensures that you capture the current day, because if you were to set the range to just one day (i.e., 5/1/2023 to 5/1/2023) the result would be zero.

3. Calculated field that computes the number of days in the range:

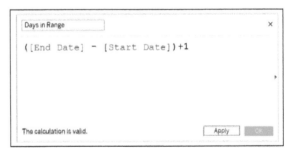

When you use parameters to filter date ranges, you don't add a traditional discrete or continuous date filter to the Filters Shelf. Therefore, to keep only the selected dates on the view, you need a calculated field that will act as the date filter. This is the Boolean formula:

[Order Date] >= [Start Date] AND [Order Date] <= [End Date]

I'm using the Sample – Superstore dataset so my date field is called Order Date, but you would replace that field with your own date field. The Start Date and End Date fields in this formula are the parameters from the first two steps earlier.

4. Calculated field that filters dates to the selected range:

The final calculated field needed is where the magic happens. This looks at the number of days in the range (computed in step 3), and based on that number, truncates the date field at the most appropriate level. For this example, let's say the following:

- When there are fewer than 31 days in the range, we'll look at *days*
- More than 30 days and fewer than 91 days in the range, we'll look at *weeks*
- More than 90 days in the range, we'll look at *months*

Here's the calculation:

```
IF [Days in Range] <= 30 THEN [Order Date]
    ELSEIF [Days in Range] <= 90 THEN DATETRUNC('week',[Order Date])
    ELSE DATETRUNC('month',[Order Date])
    END
```

Of course, you can replace the quantitative thresholds in this calculation with whatever works best for your business.

5. Calculated field that automatically truncates dates:

Now we're ready to create the view. Here's how a line graph looks with Sales on the Rows Shelf, my Automated Date Part calculated field on the Columns Shelf, and my Date Filter calculated field on the Filters Shelf. I've also shown both the Start Date and End Date parameter controls:

Note that the Date Filter is set to True, keeping only the days that match the parameter control selections.

Also note that the Automated Date Part field is set to Continuous Exact Date. We don't want to choose a specific date part because that will now be automatically controlled by the calculated field from step 5 and based on the number of days in the selected range.

Now if I choose a date range greater than 30 days but fewer than 90 days—for example, 5/1/2023 through 6/25/2023—the line graph will automatically update to look at *weeks*:

If I choose a range greater than 90 days—say, 5/1/2023 through 8/31/2023—the line graphs will automatically update to look at *months*:

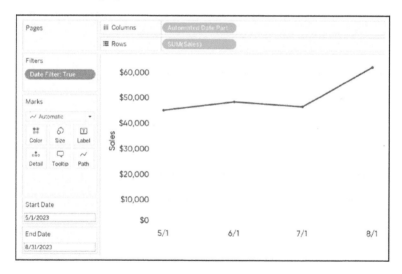

If I choose a date range containing fewer than 30 days, such as 8/1/2023 through 8/22/2023, the line graph will automatically update to look at *days*:

We've just dynamically updated the granularity of this line graph to deliver the most appropriate number of marks based on the number of days in the date range selected!

CHAPTER 69

3 Innovative Ways to Use Parameter Actions: Tip 1

There's no better topic to bridge Parts III and IV than parameters. Parameters link authors and their audiences because they allow the author to design inputs but then transfer the control of those inputs to the user. When an author can harness the power of parameters, it leads to almost infinite applications and user-friendly experiences.

When parameter actions were introduced in Tableau 2019.2, parameters became even more powerful because instead of being limited by parameter controls, the inputs can be changed when users interact with a dashboard. This chapter series shows you how to implement three of my favorite tactics with the Change Parameter dashboard action.

How to Highlight Dimension Members Using Parameter Selections

One of my favorite user experiences to provide is a drop-down box that allows the audience to choose the dimension member that is most relevant to them. I then highlight their selection on all the charts throughout the dashboard. It's like a good map that includes a You Are Here sticker to help the viewer orient themselves within the context of the rest of the visual.

The tactic is achieved through the use of parameters and a Boolean formula that identifies which dimension member matches the parameter value and which do not. For more, see Chapter 40.

Before parameter actions (explained in Chapter 2), I would have to display a parameter control to allow the user to select the dimension member to highlight. Not a terrible experience, but it was one more drop-down element taking up valuable real estate on a dashboard. This was the case with my dashboard, *Super Sample Superstore* (*https://oreil.ly/6Unk0*), in which the user can choose which region to highlight:

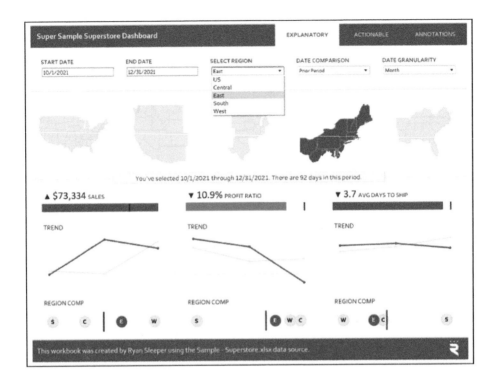

Note that the Region parameter has a data type of String and there is an allowable value for each of the four regions in the Sample – Superstore dataset. This selection highlights the dimension member on the map and dot plots, and filters all the other visuals to that specific dimension member. This experience can be made much more elegant now using parameter actions.

Any time a sheet contains the same dimension members as a parameter's allowable values, you can use the sheet as a control sheet that overwrites the parameter influencing other views. Due to this new functionality, you can create literal control sheets that are nothing but a text table containing the dimension members that you want to overwrite a parameter's value.

You can also use other visualizations as control sheets. Because every map on the *Super Sample Superstore* dashboard contains the Region dimension, I can use the maps at the top to overwrite the parameter value that is being used to highlight and filter the other views. If your control sheet does not contain the dimension members in your parameter's allowable values, simply add the dimension to the Detail Marks Card of the control sheet(s).

Here's how my Change Parameter dashboard action settings look to use my maps at the top as my control sheets:

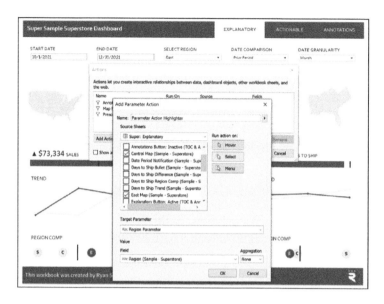

Now, simply clicking a state on the map populates the parameter value with the region that state is in. The parameter is then being used to filter and highlight all the other views. Here's how it looks after changing the dashboard's focus from East to Central by clicking anywhere in the Central region's map:

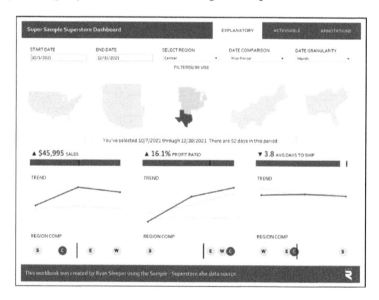

Note that in this example, I use parameters to filter the views, but I could also filter using the traditional Filter dashboard actions. However, highlighting user experience is different. This isn't like a Highlight dashboard action that makes a specific dimension member stand out, whereas everything else is faded out.

We are overwriting the allowable value in the Region Parameter. There is then a calculated field with the formula `Region = Region Parameter` placed on the Color Marks Card of all the dot plots. Whenever the Region dimension member matches the allowable value selected in the Region Parameter, it is classified as True and gets a unique color; everything else is classified as False and gets a secondary color. So, two different colors instead of a highlight; kind of a more permanent highlighter user experience.

One last note on this tip. Unfortunately, when the data type of the parameter is String, we cannot do a multiselect. Tableau will always defer to the last selection when choosing which dimension member is used to overwrite the current value of the parameter. I suppose multiselect string parameters will have to wait for a future release!

If you need to select more than one dimension member, you might be better off using *set* actions, explained in Chapter 3.

CHAPTER 70

3 Innovative Ways to Use Parameter Actions: Tip 2

Using Parameter Actions to Change Date Parts with the Click of a Button

You can create control sheets even when your dataset does not contain a dimension with dimension members that match the allowable values of the parameter you want to use to create parameter dashboard actions.

Parameters are unique in that they work across multiple data connections within the same workbook. This means that I can create a control sheet in Excel containing the dimension members that will overwrite a parameter's value, connect to that Excel sheet as a secondary data source, use it to create a control sheet, and set up a parameter action that works between the newly created control sheet and the other worksheets in the workbook.

Here's an example of why you would need to go through the extra step of creating a control sheet from a secondary data source. Another of my favorite user experiences is to provide a means for the audience to toggle the date part of a line graph from day, to week, to month, to year. However, there is no dimension that I can simply add to the Text Marks Card to create a control sheet with the different date parts. To make matters more difficult, date parts must be lowercase in order to work with certain date functions such as DATETRUNC.

If you are not familiar with this tactic, you need a parameter with a data type of String, and the allowable values are each of the four date parts. The important thing is to make the values lowercase so that they work when integrated with the second step. The Display As values can have whatever casing you prefer:

The next step is to set up a calculated field that truncates the date in the dataset at whatever level is selected in the parameter. Here's the formula for the Order Date field in the Sample – Superstore dataset:

```
DATETRUNC(
[Date Part Example],
[Order Date])
```

Now, if I create a Sales by the newly created Date with DATETRUNC calculated field, the user can control the granularity of the date via the Date Part Example parameter. For example, if I choose the Year date part from the parameter control, the line graph will update to sales by *year*.

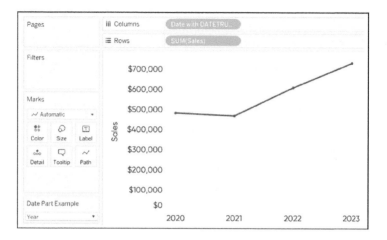

Just like our highlighter example in Chapter 69, this experience can now be improved with parameter actions and a control sheet.

First, set up a two-column Excel spreadsheet that will be used to create the control sheet in Tableau:

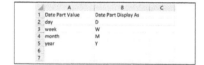

Next, connect to the data source and create a control sheet:

The critical aspect of control sheets is that the dimension containing the dimension members that will overwrite the parameter must exist somewhere on the view. In this example, I used the Display As field to make my buttons look nicer, but I placed the *Value* field on the Detail Marks Card so that it would be available as an option when I set up my parameter action later.

Now, place this control sheet on the same dashboard as the line graph and set up a Change Parameter dashboard action that will control the date part selection:

With this, simply clicking a button on the control sheet overwrites the date part in the calculated field that is controlling the truncation of the line graph. Here's how it looks after clicking W to change the date part from year to week:

The user can now decide the most appropriate date granularity for the number of marks on the view and change the date part on the fly with the click of a button!

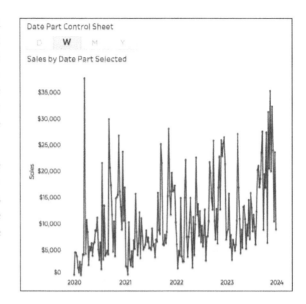

CHAPTER 71

3 Innovative Ways to Use Parameter Actions: Tip 3

Using Parameter Actions to Use Measure Names in Calculated Fields

One more of my favorite tactics is to let my users choose which measures are being displayed in a chart (see Chapter 64 from *Practical Tableau*, O'Reilly, 2018). Once again, this tactic is now even better because you can set up a parameter action that changes the measure being displayed by just clicking a dimension member on a control sheet.

First, set up a parameter containing the names of the measures from which you want to select. Here is a parameter built with all the measures that come with the Sample – Superstore dataset:

Next, set up a calculated field that gives Tableau instructions for what to display when each parameter value is selected:

Now create a chart using the newly created calculated measure. To illustrate, here I replace the Sales measure in the line graph from Chapter 70 with the new calculated

field that is controlled by the Select Measure Example parameter. If I choose Profit instead of Sales in the parameter control, the line graph updates to display the selected measure:

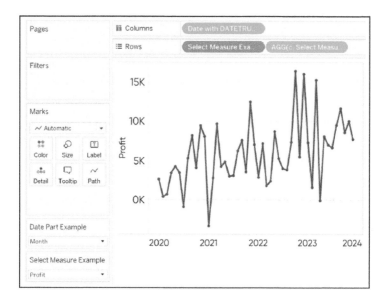

This time, for the control sheet, let's use the special generated field Measure Names. You must use Measure Names with its counterpart, Measure Values. I don't want any values to be displayed on my control sheet, but I can add the field to the Detail Marks Card to get the measure names to show up.

If there are measure names that you don't want to be part of your control sheet, simply remove them from the Measure Values Shelf. For this example, I've removed the Number of Records and Measure Selected fields, which created a filter on the Filters Shelf:

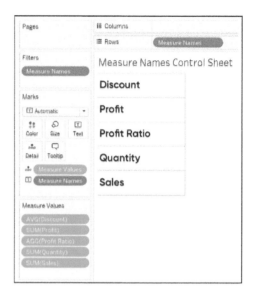

Now, we're ready to add this control sheet to the dashboard and set up the parameter action:

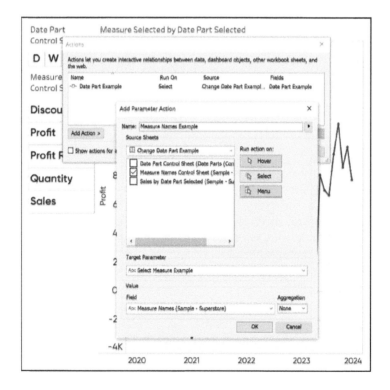

As a result, clicking a measure name from the control sheet overwrites the value in the parameter, which populates the calculated field instructing Tableau which measure to display on the graph. Here's how the view looks after clicking the Sales measure:

That's right: we just technically used the Measure Names generated field within a calculated field! That was one of the few things that was impossible in Tableau before the addition of parameter actions.

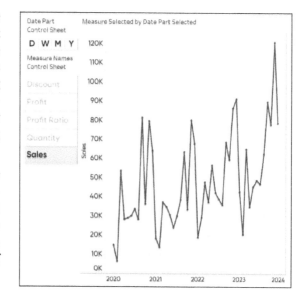

I've shown you how to use parameter actions to implement three of my favorite tactics, but there are so many ways to build from here.

One idea is to use these examples in combination with button swapping (Chapter 95) to control parameter values. The control sheets would be a series of icons instead of text, and button swapping would allow you to visually indicate which button was clicked.

These are very exciting possibilities as Tableau continues to move closer and closer to an infinitely flexible, web-like experience. Speaking of experience, keep reading on into the next section to learn ideas for improving your users' experience.

User Experience

CHAPTER 72

How to Create a Custom Top Navigation

My data visualization mission statement is to reduce the time to insight, improve the accuracy of insights, and increase engagement. Engagement is more than making pretty charts. It is also about providing an intuitive user experience to help your audience find additional value-adding insights. Part IV aims to provide several ideas and inspiration for improving your Tableau workbooks. These tips are applicable for Tableau Desktop, Tableau Public, Tableau Online, and Tableau Server.

This chapter shows you how to re-create something like the Tableau Online/Server interface within your own local Tableau workbook. This is a great technique any time you need to provide an intuitive and professional-looking interface for your end users to navigate between multiple dashboards.

Eight Characteristics of Successful User Interfaces

All of the user interface tips in this section attempt to follow the guidelines of Dmitry Fadeyev's "Eight Characteristics of Successful User Interfaces" (*https://oreil.ly/6ZXf8*):

1. Clear
It should be apparent to the end user how the interface will behave.

2. Concise
The data visualization equivalent of removing chartjunk; keep explanations succinct and to the point.

3. Familiar
Use schemas to make it easier for your users to understand what to do.

4. Responsive
The interface should react quickly/provide feedback when users interact.

5. Consistent
Coach your users through repetitive use of similar elements.

6. Attractive
Balance some design aspects to improve engagement.

7. Efficient

Design the interface in a way that makes it as efficient as possible to achieve the objective.

8. Forgiving

There should be feedback or a means of undoing mistakes made by the end user.

 Related: *Practical Tableau*, Chapter 75, "Three Ways Psychological Schemas Can Improve Your Data Visualization" (O'Reilly, 2018)

How to Create a Custom Top Navigation with Button Objects

To illustrate the following tip, let's re-create the top navigation pictured in the *Super Sample Superstore* dashboard (*https://oreil.ly/6Unk0*):

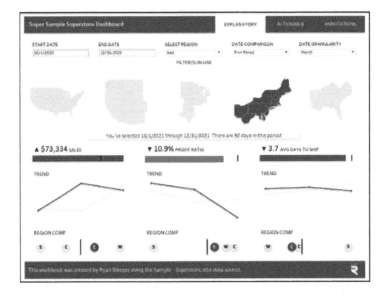

Note that the top navigation is seamlessly integrated with the rest of the dashboard with the active tab turning white, and links to the other dashboards matching the rest of the blue header. This was all created within Tableau Desktop using a series of Button (or *Navigation* as of Tableau 2020.1) dashboard objects, which I explain in Chapter 4. For each dashboard, you see an active button and two inactive buttons.

To create the first button, from the lower left of the Dashboard pane, drag a Button object onto a dashboard. This object can be floating or tiled within a layout container; I use floating Button objects so that I can change each button's dimensions and position precisely using the Layout pane:

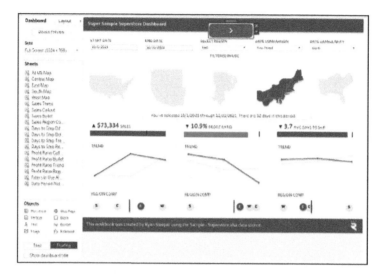

After adding a Button object, you will have a default button image, but the button doesn't link anywhere yet. To edit both the button's format and destination, click the down arrow that appears on the right side of the button when it is selected and then, on the menu that opens, click Edit Button:

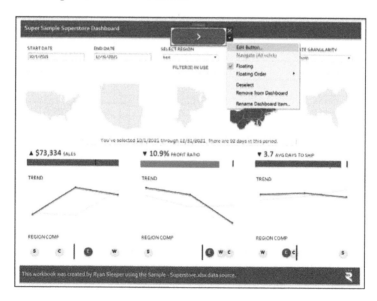

In the Edit Button dialog box, you can choose where this button should navigate within the workbook and whether you want an image button or a text button:

In this example, all of the buttons display text only, so let's choose Text Button. One nuance with text buttons is that they must link to something to get the formatting to display. So, even though the button that we use to illustrate which dashboard the user is currently viewing won't technically link anywhere, we must link it to itself to get the formatting to display properly.

Note that Button objects currently allow you to link within a workbook; to link a user outside of a workbook (i.e., to a destination on Tableau Server), use an Image object instead and set its URL.

To preview the format before you accept changes, click Apply at any time:

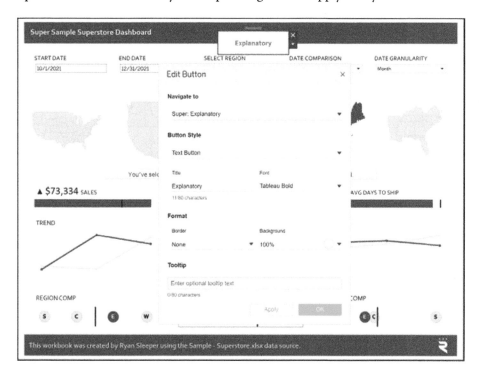

After accepting the changes, I can line up this Button object perfectly in the header by navigating to the Layout pane and changing its position and dimensions:

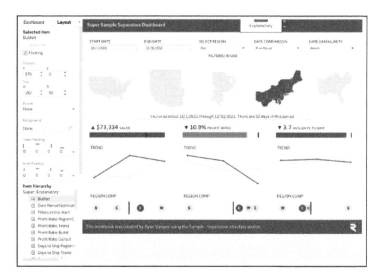

By making the background color of the button match the white on the row containing filters and parameter controls, I've created the illusion that I have a seamless navigation bar at the top of each dashboard. From here, you would just repeat these steps for each button that you want to provide. Of course, you should link each button to their respective destinations and format them as desired (i.e., in my case, inactive buttons have a blue background).

To test Button objects to ensure your user experience is working as expected, go to Presentation mode by clicking the projector icon on the toolbar above the top navigation. You can also test Button objects outside of Presentation mode by holding the Alt key in Windows or the Option key on macOS while clicking a Button object:

An interface like this works in Tableau Desktop, Tableau Reader, Tableau Public, Tableau Online, or Tableau Server!

CHAPTER 73

How to Make an Alert-Style Splash Page with Cross-Dashboard Filters

In Chapter 49 from *Practical Tableau* (O'Reilly, 2018), I show how to create a *heat map dashboard* meant to highlight data points of interest to the viewer. An alert-style dashboard like this provides the benefit of not making your end users work too hard before deciding whether they should dig deeper. If using Tableau Server or Tableau Online to subscribe to the view, they might decide that nothing needs further investigation and not interact with the dashboard at all.

Of course, if you went through the trouble of making a dashboard to monitor the performance of a particular business objective, insights are likely to emerge fairly regularly. At that point, you have a lot of control in designing the dashboard user interface in a way that helps the end user take the next step as efficiently as possible. This chapter shows you how to make an alert-style splash page, or dashboard introduction, and two ways to link into deeper analyses.

To help illustrate the following approach, you'll be using this variation of the *Super Sample Superstore* dashboard (*https://oreil.ly/6Unk0*).

In this version, the end user lands on a *splash page* that shows the performance by region and trend for each of our three KPIs: Sales, Profit Ratio, and Days to Ship (and On Time Rate). This is a very high-level explanatory view meant primarily to alert the end user when something deserves further investigation. Note that I carried over a similar menu at the top of the dashboard that we examine in Chapter 72.

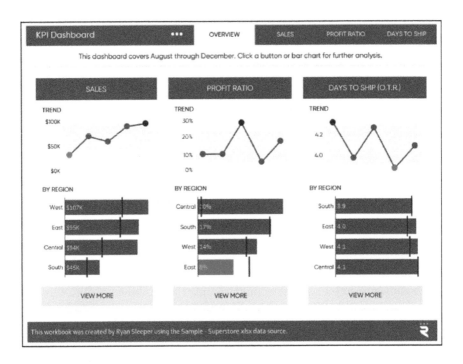

Now in addition to the top navigation, there is a button in line with each KPI widget to prompt the user to take further action. This is a slightly better user experience because they don't even need to think about which button at the top would be most appropriate to find the additional context they are looking for. You, as the dashboard author, determine what happens when a button is clicked. This means that you control the flow of the analysis even when you aren't there to walk the viewer through it.

But wait: there's more!

If I were analyzing the preceding dashboard, a few things would stand out to me. Sales is in good shape period over period across regions. There are some positive and negative anomalies in all three KPIs. Days to Ship was up across the regions, but we've maintained our on-time delivery rate. But there *is* one thing bugging me... *Profit Ratio in the East region is way down.*

If I were to click the View More button, I would be taken to the Profit Ratio dashboard, but I would need to make a few more clicks to see what's going on in the East region. Or, instead of using a Button object alone that simply navigates the user from one page to another, I can also add an action to the Profit Ratio by Region bar chart.

The advantage to adding a dashboard action is that they can be set up to both navigate the user and filter the destination dashboard; Button objects can navigate, but not filter. What I typically do when executing this tactic is to add a call to action as the tooltip of the visualization. Here's how it might look when hovering over the mark of interest:

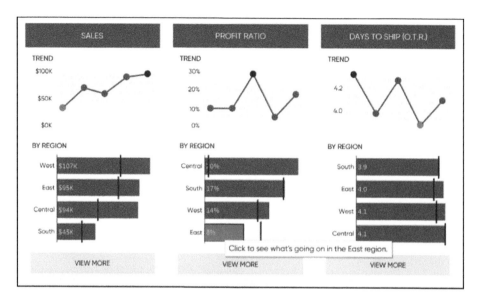

Next, to deliver on that call to action when the region is clicked, I add a Filter dashboard action that runs on Select. The trick here is to change the target sheet to the destination dashboard. For purely illustration purposes, let's set the destination to be the Explanatory dashboard in the *Super Sample Superstore* dashboard:

Note that this can be refined to filter specific sheets on the destination dashboard as well as specific fields. Now clicking the East region brings me to the Explanatory dashboard and filters it to the East region:

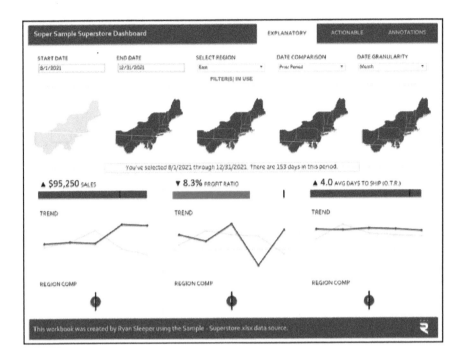

I know it's working because all maps at the top were filtered to the East region and I see only one circle in the dot plots at the bottom. This was just to illustrate the technique, but of course, you can decide where to navigate the end user and whether the action should filter or *highlight* the dimension member on some of the target sheets.

CHAPTER 74

How to Make an Intro-Style Dashboard with Cross-Dashboard Navigation

I have to admit that I have yet to use Tableau's story points feature in any of my consulting engagements. The spirit of the feature is great, and when it was released, I was excited by the prospect of replacing PowerPoint to share my data stories. But the reason I don't personally use story points stems from my point of view that getting the point across in the first screen is your best chance at causing an impact. I believe that most users simply will not click through multiple pages to figure out what you're trying to tell them.

That being said, I also believe in providing context/setup for dashboards and—as you've seen so far in Part IV—strategically breaking up views into their own dashboards and/or workbooks. This chapter shows you how to make a variety of story points to introduce a dashboard and improve the chances of your end user flipping through interior workbook pages.

Don't Bury the Workbook Lead

For one example of an intro-style dashboard, take a look at the winning entry in the 2016 War of TUGs data visualization competition, the Kansas City Tableau User Group's *Does Education Really Matter?* This workbook was meant to reaffirm that the My Brother's Keeper program is an important program to keep intact. One of several aspects I like about this entry in the contest is that although the workbook contains several pages, they didn't "bury the lead":

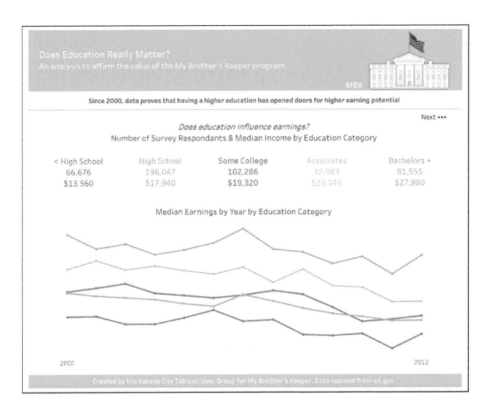

The first dashboard in the workbook asks one clear question at the top: "Does Education Really Matter?" That question is followed by one clear caption that answers the question, "Since 2000, data proves that having a higher education has opened doors for higher earning potential." Then, the visuals on the dashboard illustrate that there is a clear correlation between education-level and salary.

This introduction works because the main insight is provided right up front, which provides value on its own as a standalone dashboard. Only then, if the end user is compelled to learn more, do they have the option to use the Next button, which takes them to the next page of the story:

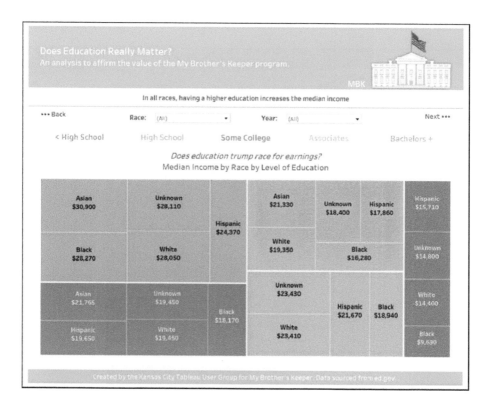

Note that the interior pages contain both Back and Next buttons, enabling the end-user to flip through the pages in both directions. We can create a similar user experience using the text buttons described in Chapter 72.

Introduction-Style Dashboard Alternative: The Welcome Mat

Another intro-style dashboard that can work well is what I call a *welcome mat*. I don't have any examples from my Tableau Public portfolio, so instead I'll give a hat tip to the first person I can remember using this technique, Matt Francis. If you're so inclined, I'm sure my friend would appreciate it if we started calling this a "Welcome Matt."

pause for laughter

Matt has several of these, so I'll just share one example:

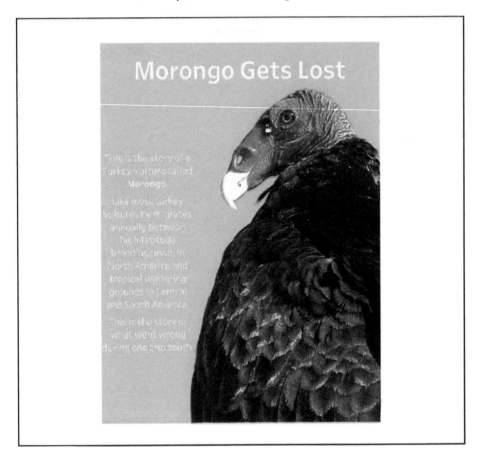

The sole purpose of this welcome mat–style dashboard is to set up the story and draw the end user in. Also note that Matt has elegantly executed a story points workbook with the navigation at the top.

Whether you are using story points or the variation offered in this chapter, I recommend that your storytelling arc follow the shape of a martini glass. This is when you offer something of value up front, walk the end user through contextual points, and then open up the analysis for self-exploration at the end. Imagine a martini glass laying on its side. The base of the glass is the main event (the foundation), the stem is the long, flat interior (the context) that connects to the mouth of the glass, which opens up to even wider than the base (the self-exploration).

CHAPTER 75

How to Add a Cross-Workbook Menu

As a consultant, I'm often asked to consolidate many views onto a single dashboard, or even many dashboards within a single workbook. Putting too much in a single dashboard or workbook has several disadvantages, including less efficient processing, increased difficulty in managing fields, and loss of focus on answering the business question at hand.

My solution for handling the too many views on a single dashboard request is to strategically group the views into separate dashboards. In Chapter 72, I illustrate how to make an integrated navigation to link multiple dashboards together.

But sometimes this is not enough. Sometimes, the separate dashboards warrant their own workbooks, and your user interface should provide an intuitive way to link those workbooks together. This chapter shows you how to use dashboard actions to add a cross-workbook menu to a Tableau dashboard.

How to Navigate Users Between Tableau Workbooks

To illustrate adding a cross-workbook menu to a Tableau dashboard, in this chapter we build on the splash page-style dashboard that Chapter 73 describes. In this version, I've added the ubiquitous three-bullet, or ellipsis, *more* icons:

Hovering over the menu provides links to other workbooks on Tableau Online, Tableau Server, or Tableau Public. I normally prefer the bullet characters approach to my cross-workbook menu design because it is efficient to make, has a subtle design, and it doesn't require purchasing an icon.

If you would like to create something similar, begin by creating a new worksheet. Next, place a dimension containing a single member onto the Text Marks Card. For efficiency, I often create a calculated dimension in my data source called Blank, with the entire formula being two quotation marks.

The reason we need this is that something has to be on the Text Marks Card before you can customize the text with the bullet characters. I prefer to use dimensions with only one dimension member to avoid having to use filters, and the Blank dimension is an easy way to accomplish this:

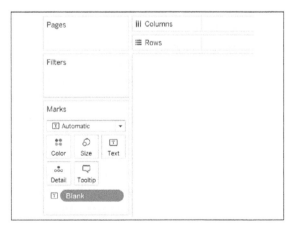

Now, we can customize the text by clicking the Text Marks Card and clicking the ellipsis (see, I told you it's ubiquitous) that appears to the right of the Text box. Here's where we paste the bullet (•) alt code three times to form my menu:

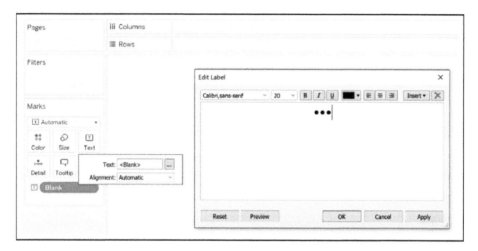

I recommend editing the tooltip to include a call to action or a description for the cross-workbook links that we will eventually provide. This will appear when the end user hovers over the menu:

Let's also format the menu by making the bullets white and changing the shading of the sheet to match the top navigation in which the menu will be placed:

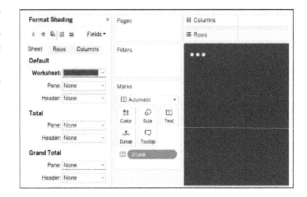

Next, go back to your dashboard where the menu will reside and simply place the sheet containing the new menu wherever you would like on the dashboard. Here, I chose to float the menu in the top navigation:

Here's where the magic happens. Now that the menu is on the dashboard, and because it is a sheet, you can add a dashboard action to it. For each link (i.e., Tableau Online, Tableau Server, or Tableau Public location), add a URL dashboard action to the menu sheet.

The most important thing is to give the dashboard action a descriptive name that will inform the end user of what they can expect to happen when they click the link. For demonstration purposes, I set up a Tableau Public link to the *Super Sample Superstore* dashboard:

Now when the end user hovers over the menu, they will see a call to action, and the links that are provided:

If the user clicks the link, the destination link that you provide in the URL dashboard action will open in a new window. Add a URL dashboard action for each link and they will all appear in the menu on hover!

You can even make a menu that looks like an icon of your choosing. If you want to take this tutorial to the next level, use the approach to create icon-based navigation or filters in Tableau from Chapter 47 in *Practical Tableau* (O'Reilly, 2018). The only difference is that you'll use a dimension with a single dimension member and add a cross-workbook link as a URL dashboard action, as described earlier.

CHAPTER 76

How to Show Top 10 Lists in Tooltips

You might have heard the recommendation to provide your dashboard users *details on demand*. Although I generally agree with the idea of a dashboard flowing from overview—to drill down (or filtering)—to specific details when needed, there can be negative consequences if your end users are too focused on the raw data. Most notably, the raw data does not provide the benefits of data visualization and often means exporting the data from Tableau—stopping the flow of thought dead in its tracks.

One of my favorite ways to provide details on demand is through Tableau's Viz in Tooltip feature, which I explain in Chapter 9. In this chapter, we set up a sheet containing the detail and add it to the tooltip of an overview or filtered visual. The challenge is that due to the order of operations of Tableau filters, it's tricky to filter the tooltip to the correct details. This chapter shows you two approaches for filtering a list to the top 10 when it's being used within a Tableau tooltip. This means that whatever dimension member you are hovering over on a dashboard will display a detailed top 10 list for that specific dimension member.

Displaying Top N Lists Using the Viz in Tooltip Feature

For this tutorial, we use a map showing profit by US state. We set up the tooltips so that when a user hovers over a state, it displays the top 10 products by profit within that state:

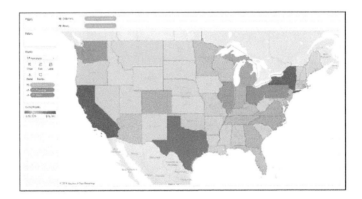

First, we need to set up a second sheet that contains the list of top 10 products. Before I share a trick for getting the top 10 list to work in the tooltips, I'm going to share a pitfall. Your first instinct is likely to set up the list as follows:

I've placed the Product Name dimension on the Rows Shelf, Profit measure on the Text Marks Card, sorted the values in descending order, and added a dimension filter that keeps the top 10 products by profit:

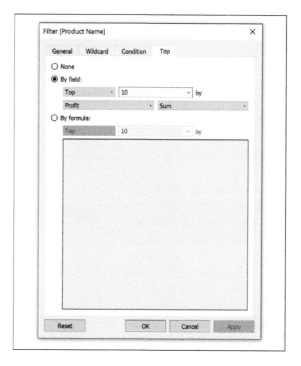

To add this sheet as a tooltip in the map, I must navigate back to the first sheet, click the Tooltip Marks Card, click the Insert button in the upper-right corner of the tooltip dialog box that appears, hover over sheet, and then choose the List sheet:

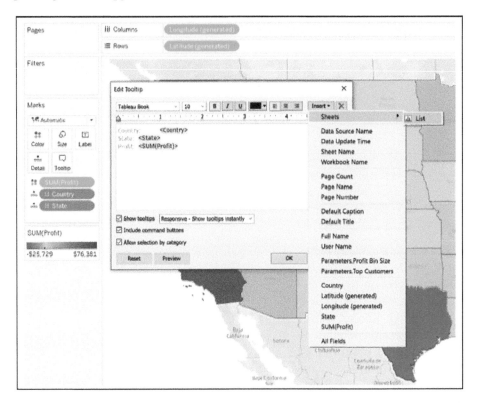

Let's also clean up the tooltip by placing the list under the other fields, adding some text to describe what's in the list, and increasing the maximum width and height of the sheet in the tooltip to 600 to ensure that the list has room to properly display:

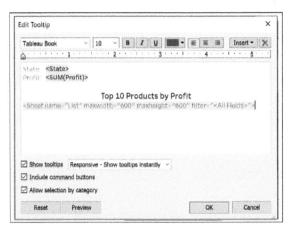

All the components are in place to display the top 10 list when I hover over an individual state, but look what happens when I hover over the state of California:

When you add a sheet to a tooltip as we've done, Tableau automatically creates a set filter to the underlying sheet. This is similar to what happens when you create a dashboard action. Here are the filters on the List sheet:

What's happening is that due to the filter order of operations, the list is being filtered to the top 10 products overall *first* and then we see the products within the overall top 10 for each state that is hovered over. We see four products when hovering over California because four of the top 10 overall were sold in the state of California at some point. If the product was never sold in a state, we wouldn't see any list when hovering over that state.

The quickest way to get the result we are looking for is to include the newly added set filter to context so that the list filters to each state first, followed by the top 10 filter for that state second. To add a filter to context so that it moves up in the order of operations, on the Filters Shelf, click the filter and then choose Add to Context. The filter will turn gray to indicate that it has been added to context:

Now when we hover over the state of California, or any other state, the list is filtered to the state first, and then the top 10 for that state is displayed in the tooltip:

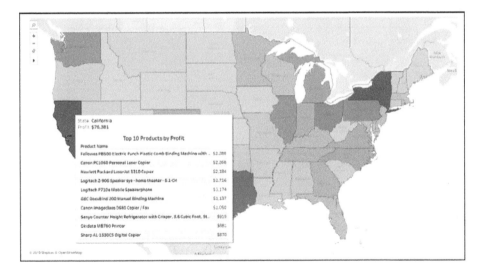

There's a slightly more clever solution that will give you more flexibility to show the top or bottom 10 that involves using the INDEX() function for top *N* filtering (see Chapter 58).

First, set up a calculated field with the formula INDEX() <= 10 (or whatever number you want to use as the top or bottom *N*):

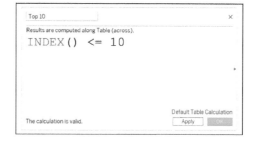

INDEX() is a table calculation that returns the row number, so this is a Boolean formula that will keep only the first 10 rows on the view. Let's replace the Product Name filter with this newly created calculated field and choose True (keeping only the first 10 rows). With this second technique, you no longer need the set filter added to context, so we can also remove that filter from context by clicking the gray filter and choosing Remove from Context.

Note that I only need to remove the filter from context because I'm building on my first example. If you opted to start with this second approach, the second filter would never have been added to context to begin with:

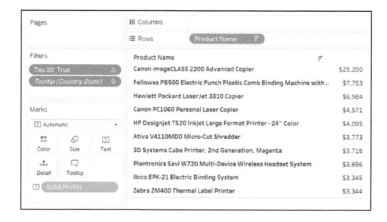

This approach will always keep the first 10 rows displayed regardless of which state is hovered over. The only catch with this approach is that you must ensure the list is sorted appropriately. If you want to show the top 10 products in this case, we would need to make sure the list was sorted in *descending* order. What's nice about this is that we can easily display the bottom 10 by simply flipping the sort order from descending to ascending. This puts the product name with the lowest profit value first, and our Boolean filter keeps the 10 worst performers on the view:

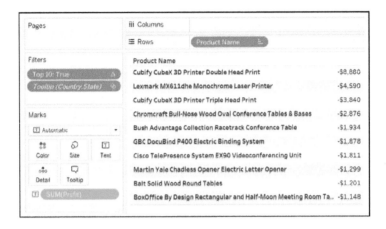

Pages			iii Columns	
			≡ Rows	Product Name ⌐
Filters			**Product Name**	
Top 10: True		⌐	Cubify CubeX 3D Printer Double Head Print	-$8,880
Tooltip (Country,State)		⌐	Lexmark MX611dhe Monochrome Laser Printer	-$4,590
			Cubify CubeX 3D Printer Triple Head Print	-$3,840
Marks			Chromcraft Bull-Nose Wood Oval Conference Tables & Bases	-$2,876
⊡ Automatic		▾	Bush Advantage Collection Racetrack Conference Table	-$1,934
⇅	⊘	⊤	GBC DocuBind P400 Electric Binding System	-$1,878
Color	Size	Text	Cisco TelePresence System EX90 Videoconferencing Unit	-$1,811
∘∘∘	▢		Martin Yale Chadless Opener Electric Letter Opener	-$1,299
Detail	Tooltip		Balt Solid Wood Round Tables	-$1,201
⊡ SUM(Profit)			BoxOffice By Design Rectangular and Half-Moon Meeting Room Ta..	-$1,148

In our map view, Texas is interesting because it appears to be the state in which we lost the most profit. Now, if I hover over Texas, the top 10 list will be shown; only now it's been toggled to display our bottom 10 products by profit (I've also changed the list title accordingly):

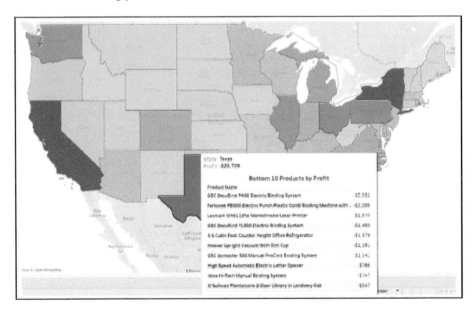

Note that throughout these examples I've used top 10 and bottom 10, but of course you can set these up to show whatever top N or bottom N is appropriate for your business.

How about showing both the top N and bottom N on the same list?

I'll close this chapter with one last idea. You can set up two separate sheets—one sorted in descending order to keep the top *N*, and one sorted in ascending order to keep the bottom *N*—and add them both to the tooltip.

But you can also consolidate the two sheets into one by modifying your `INDEX() <= 10` formula to the following:

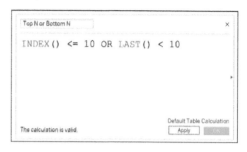

```
INDEX() <= 10 OR LAST() < 10
```

`LAST()` is another table calculation that computes the difference between the current row and the last row. A difference of less than 10 would represent the last 10 rows. The difference between the last row and itself is 0, the second-to-last row is a difference of 1, and the difference with the tenth-to-last row is 9. When your table is sorted in descending order, this formula would keep both the top 10 and bottom 10 on the same table!

CHAPTER 77

How to Color the Highest and Lowest Points on a Highlight Table

When I'm educating my stakeholders on the value of data visualization, I begin by showing a raw crosstab of data—similar to what most corporate reports still look like today—and ask the audience to answer the basic business question of identifying the highest or lowest number in the table. I then convert the crosstab to a highlight table by introducing the preattentive attribute of color, which reduces the time to insight, increases the accuracy of insights, and improves engagement.

In the exercise, I take the highlight table a step further by coloring only the highest and lowest number in the view, further reducing the time to insight and increasing the accuracy of insights. This chapter shows you how to highlight the highest data point and lowest data point on a view using table calculations. This is as much about sharing some technical know-how as it is about introducing the important concept of *using Tableau to automatically answer business questions* for you and your audience.

How to Use Table Calculations to Automatically Answer Business Questions

To illustrate how to highlight the maximum and minimum values on a visualization built in Tableau, let's begin with this crosstab showing sales broken down by the Sub-Category and Month of Order Date dimensions:

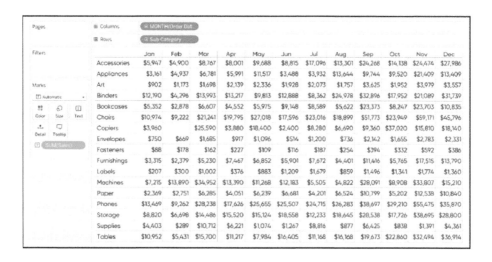

Sub-Category	Jan	Feb	Mar	Apr	May	Jun	Jul	Aug	Sep	Oct	Nov	Dec
Accessories	$5,947	$4,900	$8,767	$8,001	$9,688	$8,815	$17,096	$13,301	$24,268	$14,138	$24,474	$27,986
Appliances	$3,161	$4,937	$6,781	$5,991	$11,517	$3,488	$3,932	$13,644	$9,744	$9,520	$21,409	$13,409
Art	$902	$1,173	$1,698	$2,139	$2,336	$1,928	$2,073	$1,757	$3,625	$1,952	$3,979	$3,557
Binders	$12,190	$4,296	$13,993	$13,217	$9,813	$12,888	$8,362	$24,978	$32,896	$17,952	$21,089	$31,739
Bookcases	$5,352	$2,878	$6,607	$4,552	$5,975	$9,148	$8,589	$5,622	$23,373	$8,247	$23,703	$10,835
Chairs	$10,974	$9,222	$21,241	$19,795	$27,018	$17,596	$23,016	$18,899	$51,773	$23,949	$59,171	$45,796
Copiers	$3,960		$25,590	$3,880	$18,400	$2,400	$8,280	$6,690	$9,360	$37,020	$15,810	$18,140
Envelopes	$750	$669	$1,685	$917	$1,096	$514	$1,200	$736	$2,142	$1,655	$2,783	$2,331
Fasteners	$88	$178	$162	$227	$109	$116	$187	$254	$394	$332	$592	$386
Furnishings	$3,315	$2,379	$5,230	$7,467	$6,852	$5,901	$7,672	$4,401	$11,416	$5,765	$17,515	$13,790
Labels	$207	$300	$1,002	$376	$883	$1,209	$1,679	$859	$1,496	$1,341	$1,774	$1,360
Machines	$7,215	$13,890	$34,952	$13,390	$11,268	$12,183	$5,505	$4,822	$28,091	$8,908	$33,807	$15,210
Paper	$2,369	$2,751	$6,285	$4,051	$6,239	$6,681	$4,201	$6,524	$10,799	$5,202	$12,538	$10,840
Phones	$13,469	$9,262	$28,238	$17,626	$25,655	$25,507	$24,715	$26,283	$38,697	$29,210	$55,475	$35,870
Storage	$8,820	$6,698	$14,486	$15,520	$15,124	$18,558	$12,233	$18,645	$28,538	$17,726	$38,695	$28,800
Supplies	$4,403	$289	$10,712	$6,221	$1,074	$1,267	$8,816	$877	$6,425	$838	$1,391	$4,361
Tables	$10,952	$5,431	$15,700	$11,217	$7,984	$16,405	$11,168	$16,168	$19,673	$22,860	$32,494	$36,914

Without the preattentive attribute of color, it is challenging to quickly and accurately answer the question at hand: *what is the highest and lowest value in the table?* To help us, let's create a calculated field with the following code, which will automatically answer the question for us:

```
IF SUM([Sales]) = WINDOW_MAX(SUM([Sales])) THEN "Max"
ELSEIF SUM([Sales]) = WINDOW_MIN(SUM([Sales])) THEN "Min"
ELSE "Neither"
END
```

What this calculation will eventually do is look at the value of each cell, and if that value matches the maximum value in the table, it will be labeled *Max*; if the value matches the minimum value in the table, it will be labeled *Min*. To have Tableau help us answer the business question, we are using two table calculations: WINDOW_MAX and WINDOW_MIN. By default, Tableau table calculations are computed from left to right across the table. This means that by default, the highest and lowest numbers will be labeled for every row in the table.

Here's how the view looks when we add our new MIN / MAX highlight calculated field to the Color Marks Card:

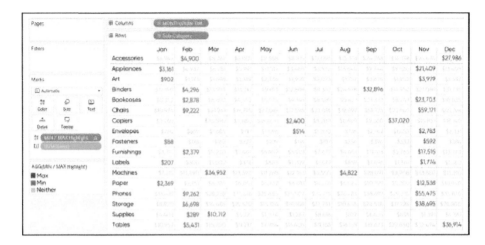

To get the effect we are hoping for, we will need to change the addressing of the table calculation. Currently, the highlighter is being computed with Table (Across), or Month of Order Date, only. To highlight the highest and lowest values for the entire table, we need to change the addressing to Month of Order Date *and* Sub-Category.

There are at least two ways we can change this. First, from within the view, you can click into the pill with the table calculation (identified with a delta symbol) and then click Edit Table Calculation. In the dialog box that opens, choose Specific Dimensions, and then select the checkboxes for both Month of Order Date and Sub-Category:

You can get to the same
answer by clicking into the pill
with the table calculation, hovering over Compute Using,
and choosing Table (across
then down):

In both cases, the highest number in the entire table is rendered in one color, the lowest number in the table is a second color, and everything else appears in a third color:

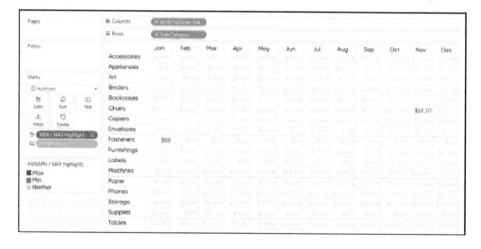

To conclude, let me reiterate a little hack from Chapter 18 to create a highlight table,
even when the coloring is being created by a discrete dimension (like the MIN / MAX
highlighter created here) instead of a continuous measure (like you would see on a

traditional highlight table with a measure like Sales). Begin by changing the mark type to Square and maximizing the size of the squares using the Size Marks Card:

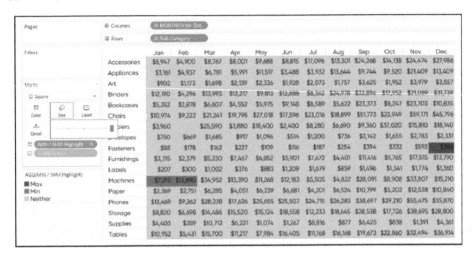

This view is not ideal because the squares are bleeding over into multiple cells. The trick to containing the highlighted squares in the correct cells is to add a Blank dimension to both the Columns Shelf and Rows Shelf. To do this, double-click to the right of each dimension that is currently on the Columns Shelf and Rows Shelf, type two quotation marks, and press Enter.

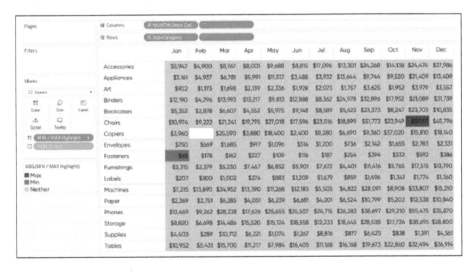

If the extra space that the blank headers are creating bothers you, just right-click each header and deselect Show Header.

Note that I used this highlight table example to illustrate that the logic was working properly, but you can use this type of effect for any chart type. We have created a calculated field that automatically answers a business question for us! This approach has reduced the time to insight and increased the accuracy of those insights even further than the preattentive attribute of color could do on its own.

CHAPTER 78

How to Automatically Highlight the Latest Trends

This tutorial illustrates how to add a highlighter to the most recent data point on a line graph. You also can dynamically color this highlighter to communicate that the latest trend was either positive or negative. This technique provides several benefits, such as orienting your end users, communicating positive and negative performance changes at a glance, and enhancing the dashboard design.

To highlight the latest trend, let's use the trick from Chapter 63 in *Practical Tableau* (O'Reilly, 2018), "How to Highlight a Dimension." Even though we can add reference lines to dates, we will need to get a little hackier to produce the effect we're looking for. In this exercise, we combine this trick with a calculated field to automatically turn on the highlighter for the most recent month in our analysis and color it to communicate a positive or negative change.

First, I'm going to lay the foundation for the chart by making a Sales by continuous Month line graph from the Sample – Superstore dataset:

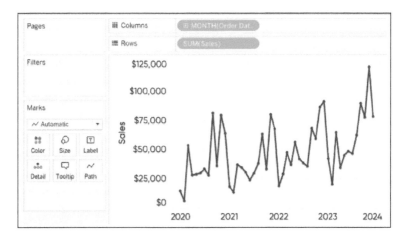

Next, let's create a calculated field that will eventually draw a bar for the most recent month being visualized. Here's the formula:

```
IF ATTR(DATETRUNC(
'month',[Order Date])) =
WINDOW_MAX(ATTR(DATETRUNC(
'month',[Order Date])))
THEN 1
END
```

It looks like we're getting a little complex, but this formula simply causes Tableau to look at each month in the line graph and determine whether it matches the latest month. When it matches, it shows the number 1, which will be used in the next step. Because this is a table calculation, I had to use the ATTR function to aggregate the dates.

Next, put the newly created Max Month calculated field on the Rows Shelf:

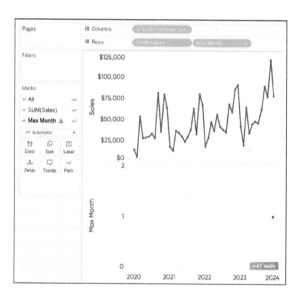

When there are two continuous measures on the same shelf, you can combine them into a dual-axis combination chart. Yes, this is application #238 or so for this feature. See Chapters 19 through 21 for just a few other uses of this chart type.

After converting the view to a dual-axis chart, you can independently change the mark type of the measures. For this example, let's change the mark type for Max Month to Bar:

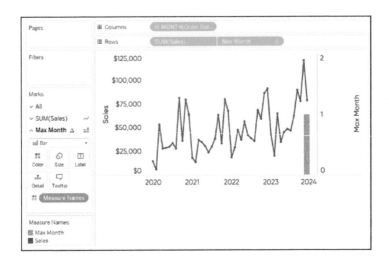

At this point, the bar will always show up on the latest month in the view, even when the dates are filtered or if you're using a relative date filter to always display data through the latest full month. Before we add color, I'd like to make a couple of formatting changes. First, fix the right axis from 0 to 1 by right-clicking the axis. Click Edit Axis on the menu that opens and fix the start and end. Make the bar evenly distributed around the latest point on the line graph by clicking its Size Marks Card, choosing Manual, and then increasing the width of the bar. Finally, let's increase the transparency of the bar by clicking the Color Marks Card and sliding the opacity slider to the left:

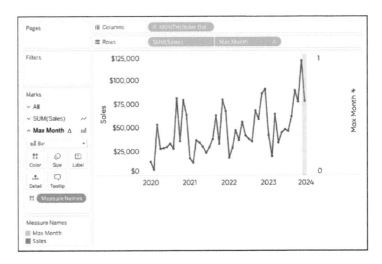

To color the bar, on the Color Marks Card for the bar, place the measure being evaluated (in this case Sales). Then, add a Quick Table Calculation that computes the month-over-month difference in sales:

This will work as is, but I usually change the color mapping so that anything positive is one color and anything negative is another color. To do this, double-click the color legend, change the steps to 2, and force the center to be 0 (under the advanced options). The last thing I do is hide the right axis for Max Month by right-clicking the axis and then deselecting Show Header.

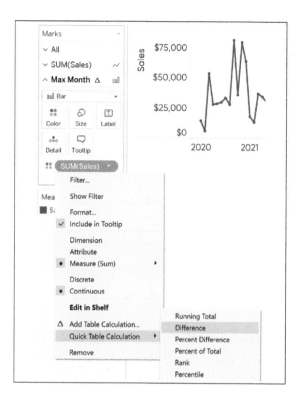

Now, whenever the most recent month had a negative month-over-month performance, a red highlighter will display:

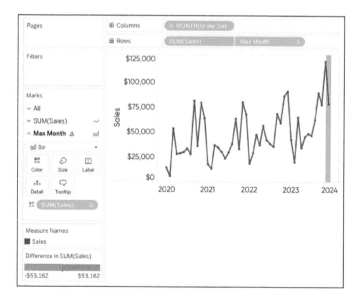

But even if the view is filtered to an older month or our data is refreshed next month, the highlighter will automatically be displayed on the latest data point and colored based on month-over-month performance. Here's an example when I filtered the dates through the latest August in the dataset:

We've not only improved the design of this traditional line graph, but the new design element also provides the practical purpose of indicating whether performance improved or declined!

CHAPTER 79

How to Do Anomaly Detection

I think we've all been there: we are sharing a meaningful story we found in the data, only to have our end users fixate on a previous peak or valley in our visualization. This often derails the conversation at hand or prevents our audience from hearing the rest of our message or reduces our chances of causing action.

One of the biggest challenges we face as data visualization practitioners is helping our end users to avoid distraction. When our end users become distracted, it makes it more challenging to communicate the story in the data and our recommended actions. Ironically, one of the reasons users become distracted is that visualizing data makes it much easier to spot points of interest. Unfortunately, just because something might pique interest, it is not always relevant to the conversation.

This chapter shares an approach to doing anomaly detection in Tableau. With anomaly detection, you're able to focus on the data points that matter and have a statistical explanation for your end users to help avoid distracting conversations.

Using Table Calculations to Do Statistical Anomaly Detection in Tableau

To help illustrate an approach for doing anomaly detection in Tableau, we re-create the sales-by-month trend using the Sample – Superstore dataset. Note that the reference distributions show whether each data point is within one or two standard deviations from the mean and the circles are colored based on whether they are an anomaly. For the purposes of this exercise, *anomalies* are defined as being at least one standard deviation from the mean:

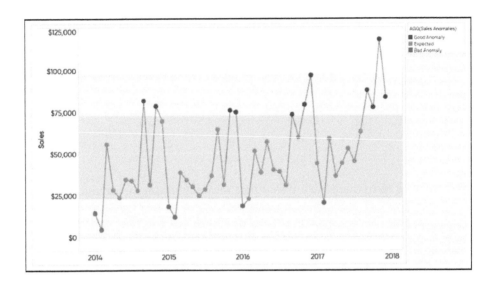

To begin, create a line graph with your measure of interest:

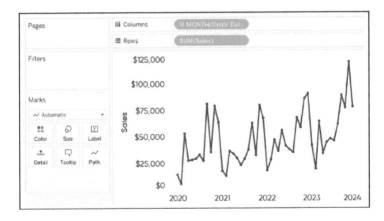

Next, add reference lines to shade the distribution between –2 to 2 standard deviations and –1 to 1 standard deviations. Note that if you want to display more than one distribution, it is best to begin with the widest distribution first so that your overlapping colors work.

To add a reference distribution, I like to right-click the axis and choose Add Reference Line on the menu that opens, click the Distribution tab, change the value to Standard Deviation, and put in my factors (in this first case, –2,2).

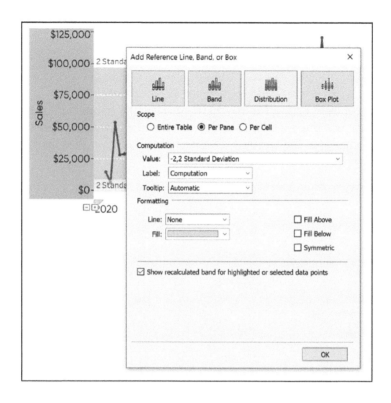

Optionally, use a different reference distribution, add a second range, and/or format the reference distributions to your liking. Here's how the view looks after we add a reference distribution for –2,2 standard deviations, add a reference distribution for –1,1 standard deviations, and clean up the formatting:

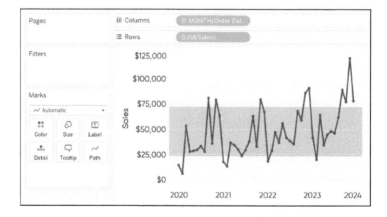

Related: *Practical Tableau*, Chapter 43, "How to Make a Control Chart" (O'Reilly, 2018)

At this point, we have a basic control chart. To create the colored anomaly indicators, let's use a dual axis and a table calculation to color the circles. To start the dual-axis combination chart, put the measure of interest onto the Rows Shelf for a second time and change the mark type to Circle:

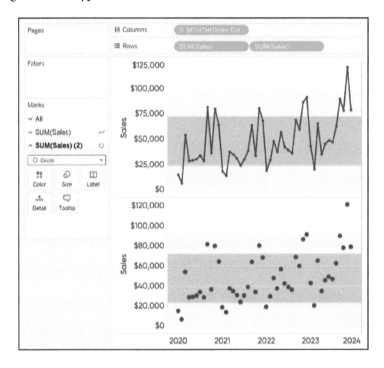

Next, convert the view into a dual-axis chart: on the Rows Shelf, right-click the second occurrence of the pill and then choose Dual Axis. Also, ensure that the axes are lined up by right-clicking either axis and choosing Synchronize Axis. At this point, I also like to hide the right axis by right-clicking it and deselecting Show Header:

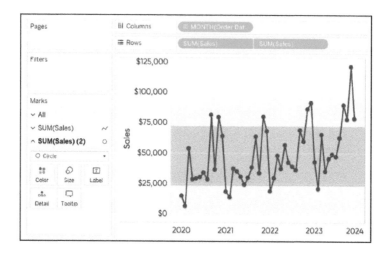

Now that there are two pills on the Rows Shelf, they each have their own set of Marks Cards that we can independently edit. This means that we can change or keep the color of the line one thing, but color the circles by whether they are considered an anomaly.

As I mentioned earlier, I'm defining anomalies as any data point outside of one standard deviation from the mean. The formula for this definition is as follows:

```
IF SUM([Sales]) < (WINDOW_AVG(SUM([Sales])) - WINDOW_STDEV(SUM([Sales])))
THEN "Bad Anomaly"
ELSEIF SUM([Sales]) > (WINDOW_AVG(SUM([Sales])) + WINDOW_STDEV(SUM([Sales])))
THEN "Good Anomaly"
ELSE "Expected"
END
```

This formula looks a little intimidating, but all it's doing is looking at each data point at the most granular level of detail (i.e., month) and comparing that number to the mean across the view, plus or minus one standard deviation. If the value is not greater than one standard deviation from the mean or less than one standard deviation from the mean, it is classified as Expected.

I'm using Sales as my measure, but simply replace Sales with the measure of your choice if you want to use something else. Also, if you'd prefer to use 2, 3 (or even 1.5), and so on standard deviations, just multiply WINDOW_STDEV(SUM([Sales]))) by 2, 3, or 1.5, respectively.

After you have the calculated field for anomaly detection, place it on the Color Marks Card for the circles:

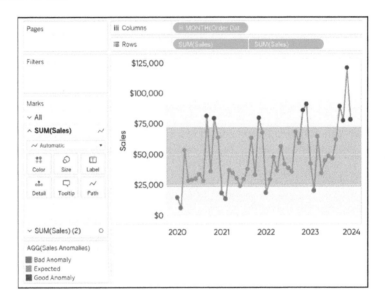

This makes the following possible:

- You can see meaningful insights very efficiently and accurately.
- You know when you shouldn't get too excited about a spike (and when you should).
- You can help your end users avoid becoming distracted when dips are not significant.

For example, in this view built with the Sample – Superstore dataset, I can see that the last four months in a row have been positive anomalies. Also, I see that even though the last drop looks pretty steep compared to the previous month, it is still greater than one standard deviation than the four-year mean displayed. So, we've taken a potentially negative distraction, and positioned it as the positive story that it is.

CHAPTER 80

How to Highlight Data Points in a Custom Date Range

This tutorial illustrates how to add a highlighter to a custom date range selected by an end user. The highlighter lies over the data points corresponding with the selected range of dates. This technique helps you and your dashboard's users to see the marks in the selected date range in context of the other marks on the view and is an easy way to enhance the design of a line graph. One of the best parts about the following approach is that it does not require you to use the dual axis, leaving some flexibility to add even more context or aesthetic improvements to the graph.

To highlight the data points in a custom date range, let's use a combination of Tableau parameters and reference lines. Just as with Chapters 78 and 79, let's create a *sales by continuous month* line graph using the Sample – Superstore dataset as our foundation:

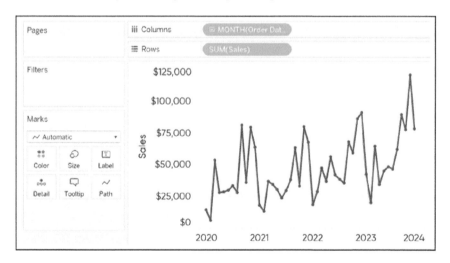

Next, set up a parameter with a data type of Date. This first parameter will eventually be used to select the beginning of the highlighted range. Here's how it looks after setting it up and choosing 7/1/2023 as the date:

We also need to create a parameter that we can use to select the end of the highlighted date range. Because we already made a parameter with a data type of Date, you can either create a new parameter from scratch or simply right-click the Start Date parameter created in the previous step and then, on the menu that opens, choose Duplicate.

Either way, after we edit the name of the parameter to End Date and choose a different date (9/30/2023), here's what the second parameter looks like:

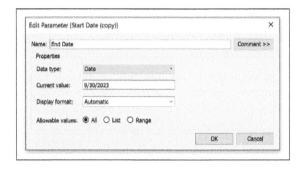

After creating both parameters, we can create a highlighter that will surround the data points in the specified date range. This is accomplished through a reference line, or more specifically, a reference *band*. To add a reference line, you can either use the Analytics pane or just right-click the axis and choose Add Reference Line.

There are just three keys to get this to work properly:

1. At the top of the Add Reference Line, Band, or Box dialog box, click the Band icon.

2. In the Band From section, click the Value drop-down box and change it to the newly created Start Date parameter.

3. In the Band To section, click the Value drop-down box and change it to the newly created End Date parameter.

Here is where you might also choose to format the band and its enclosing lines or customize the line labels (I like to change them to None). Here's how the reference band looks after setting it up and tweaking the format:

You can already see the reference band working in the background as it's highlighting the dates set up in the parameters (July 1 through September 30). Click OK to close the dialog box.

Lastly, show both the Start Date and End Date parameter controls by right-clicking each parameter and selecting the Show Parameter Control checkbox. This gives you and your end users one-click access to change either the start date and/or end date of the highlighter. Here's how the final view looks after highlighting the year 2022:

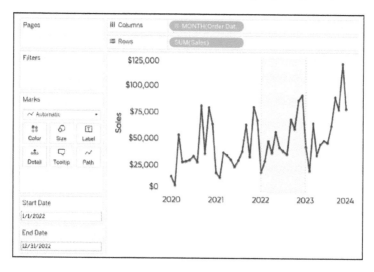

This highlight effect will always update to show the selected date range in context of the other marks on the view! Because we used reference lines instead of a dual-axis combination chart to create the effect, we still have the second axis available to add value in a different way. For example, we could add anomaly detection, as described in Chapter 79.

CHAPTER 81

How to Add a Date Period Notification to a Dashboard

When you show a date filter or the parameter controls for a start date and an end date, the end user can see what dates are being shown on the view. However, these filters and controls are not always shown to the end user (if you have a relative date filter showing the last 30 days, for example). Even when the dates are visible, in my experience it's not always apparent to end users which date range is included.

This tutorial shows you how to add a date range notification to a Tableau dashboard. The following approach not only alerts end users that the date range is being filtered, but also communicates what date range is being displayed and how many days are in the selected period. The result looks like the yellow alert on the *Super Sample Superstore* dashboard (*https://oreil.ly/6Unk0*). Note the highlighted band communicating the date range being used as well as the period in days.

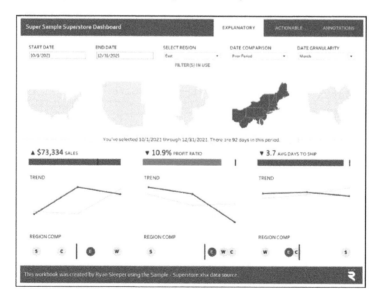

In this chapter, I show you how to do this when two parameters are being used to select a date range (one for start date and one for end date). This is often my preferred approach to filtering dates because we can use the parameters in so many ways for calculations and comparisons. The steps are very similar if you prefer to use a continuous date filter. To begin, start a new worksheet.

There are a couple of calculations we need. First, a Boolean formula that keeps only dates on or after the Start Date and on or before the End Date. Here's the formula:

```
DATE([Order Date]) >= [Start Date] AND Date([Order Date]) <= [End Date]
```

Place the newly created calculated field on the Filters Shelf and keep only the True values. To test it and lay the foundation for our notification, place the Date field being used in your formula onto the Text Marks Card twice; once with an aggregation of MIN and once with an aggregation of MAX. Here's how the view looks so far:

I can see that it's working so far because the MIN date and MAX date match the Start Date parameter (10/1/2021) and the End Date parameter (12/31/2021).

The next calculation is optional, but if you also want to compute the number of days in the selected time period, create a calculated field with this formula:

```
DATEDIFF(
'day',[Start Date],
[End Date])+1
```

The +1 at the end of the formula ensures that the last day in the period is included in the count. You can also quality check this formula by placing it on the Text Marks Card; just be sure to change the aggregation to Minimum, Maximum, Average, Median, or use Attribute to deduplicate the count and get the correct answer:

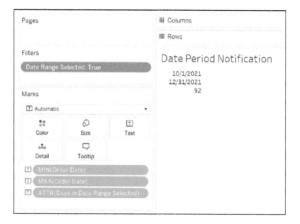

When all of the ingredients for the notification are on the view, we're ready to format the text. To do this, click the Text Marks Card and then click the ellipsis directly to the right of the Text box. Here, you can reorder the fields, change the font, and add hardcoded words:

To finalize the view, let's add some formatting to give it a highlight look and feel. These options are all up to you, and you can find them by right-clicking anywhere in the view and then choosing Format. Here's how the completed sheet looks:

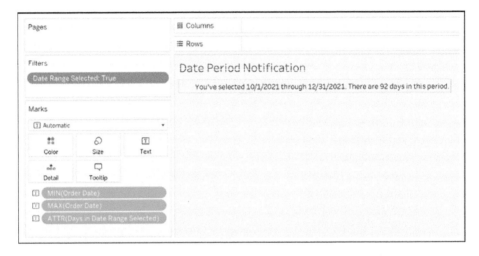

Lastly, add this to the dashboard(s) as you want, and you have a date period notification that will update whenever a change is made to the date filters!

CHAPTER 82

How to Turn Data Normalization On and Off

Data normalization is the process of adjusting values from different scales to a common scale, providing a better apples-to-apples comparison of the values. For example, looking at varying metrics like "total cheeseburgers eaten per year by US state," "total high fives given per year by US state," and "total vacation days taken per year by US state," will likely show that California leads the way in all three categories. This makes it sound like California is the best place to be in the US, and maybe it is, but these results are most likely due to California having the highest population (i.e., more people around to give each other high fives).

A better analysis would be to determine how many high fives are given *per person* per year by US state. There are several ways to normalize data in Tableau including changing the aggregation of a measure, creating a calculated 100-point index, or setting a common baseline. There are also times when it's valuable to see the raw, unnormalized numbers so that our end users have the context of the original scales.

This chapter shares a method for normalizing data and providing you and your end users the ability to toggle the normalization on and off.

How to Toggle Between Actual and Normalized Data

As mentioned, there are several methods for normalizing data in Tableau. For this tutorial, we provide a means for our end users to toggle a metric between Sales by US State (unnormalized) and Sales Per Person by US State (normalized). If you are using a different method to normalize your data, you should be able to use the following approach but tweak the logic to match your method.

As usual, we use the Sample – Superstore dataset, but that data source does not have a metric for population. So, the first step in this case is to append the metric required to normalize the data to the data source.

To do this, I looked up the population per state, made a simple two-column data source (State, Population), and joined it to the Sample – Superstore dataset:

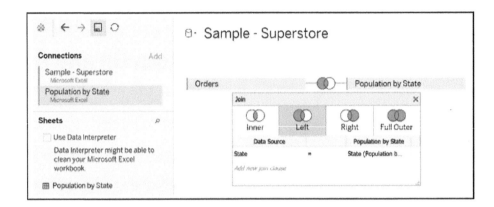

One quick aside, but when you join tables in this manner, it's common for the rows in the table on the right to repeat because they'll likely have more than one match to the table on the left. You can easily alleviate this by creating a calculated field with a FIXED level of detail expression that brings

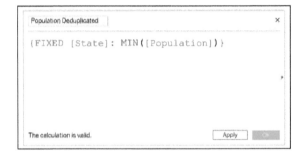

back only one row at a time. Here's the formula in this case:

    {FIXED [State]: MIN([Population])}

For more on this approach, review Chapter 50.

The method for toggling normalization on and off involves creating a parameter. Now that we have the metric we need to normalize the sales data, we must create a parameter by right-clicking any blank space in the Data pane and then, on the menu that opens, choosing Create Parameter.

For this case, I'm going to set the data type to Boolean because there are only two possible choices: normalization on and normalization off.

I give the two choices friendly names to help my end user make their preferred choice:

After we set up the parameter, we need to give Tableau instructions for what to do when each selection is made. When normalization is turned on, we want to divide the metric being normalized by the population metric that we created earlier. Here, we're using population, but you can tweak the logic and use your own denominator as desired. When normalization is turned off, we show the original value. To do this, let's create a new calculated field; here's how the formula looks when normalizing the Sales metric:

```
IF [Normalization Toggle] = True THEN SUM([Sales]) / SUM([Population Deduplicated])
ELSEIF [Normalization Toggle] = False THEN SUM([Sales])
END
```

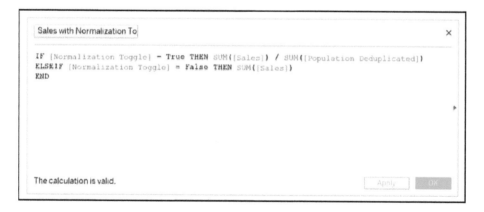

Lastly, if you want you and your end users to have one-click access to toggle the normalization on and off, ensure that you show the parameter control. To do this, in the Parameters area of the Data Pane, right-click the parameter and then choose Show Parameter Control.

Here's how the final view looks with a filled map for sales by state. In the first image, the Sales measure is not normalized, showing that the most-populated states lead the way in absolute sales volumes.

Sales by US State—Not Normalized:

However, when we toggle the normalization on, we see a very different story!

Sales by State—Normalized:

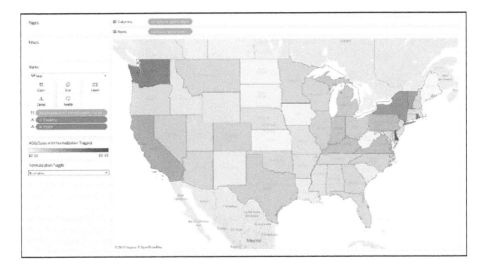

CHAPTER 83

How to Add a Filter(s) in Use Alert to a Dashboard

As I am constantly touting the benefits of data visualization and Tableau, I regularly have conversations with the objective of helping analysts transition their reporting from spreadsheets and Excel. It is not that I dislike Excel (though I can't say the same about spreadsheets as *data visualizations*). I use Excel almost daily, and it is excellent software for preparing and storing data sources that require fewer than Excel's roughly one-million-row limit.

Being such a pervasive tool and pioneer in office software, Excel is bound to have a few valuable features that haven't yet made it into Tableau. Remember when the XFL had the skycam before the NFL made it cool? Orlando Rage—good times. Anyway, one of those features that I really like about Excel that is missing in Tableau is an alert that informs you when a filter is being applied to a dashboard. In fact, Excel tries to give you a heads up in several ways, including:

- A filter icon on the column being filtered
- A caption in the lower-left corner of the view
- Highlighted row numbers

In Tableau, unless a filter is being shown on a dashboard, it's possible to have *no indicators* to an end user that the view is being filtered. Even when a filter is being shown on a dashboard, the indication that it's in use is very subtle (i.e., showing the filter selection). When you're jumping around between explanatory and actionable views or if your end user is not experienced in Tableau, it can be easy to lose track of when a filter is being applied. This chapter shows you a simple four-step solution for adding a Filter(s) in Use alert to a Tableau dashboard.

How to Use Level of Detail Expressions to Know When a Dataset Is Being Filtered

For this exercise, we use the *Super Sample Superstore* dashboard (*https://oreil.ly/6Unk0*) to illustrate this approach. As you can see from the following screenshot, when we filter down to a region other than US (which is the unfiltered view), a caption appears alerting the end user (look just beneath the Region filter):

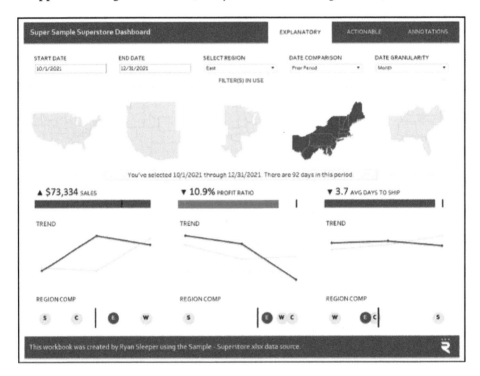

The first step to creating a Filter(s) in Use alert is to isolate the total number of records in your dataset. The resulting number is used to calculate whether a filter is being applied. If the number of records in the view matches the number of records in the entire dataset, no filter is being applied. If the number of records in the view is fewer than the number of records in the entire dataset, a filter is being applied.

To compute the unfiltered number of records in the dataset, create a calculated field that uses a FIXED LOD expression. Here's the formula:

```
{FIXED: SUM(
[Number of Records])}
```

Note that Number of Records is a generated field that will automatically be available for use no matter what data source you are using.

You can also write this formula as `{SUM([Number of Records])}`, but I want to explicitly show you that a FIXED LOD expression is taking place.

This formula instructs Tableau to count the total number of records in the dataset regardless of the filters being applied. One caveat to this is that the record count happens after extract filters, data source filters, and context filters.

The next step is to create a second calculation that computes whether the number of records in the view matches the unfiltered number of records. When the two numbers *do not* match, I want a FILTER(S) IN USE message to show. When the two numbers *do* match, I don't want an alert.

Here's the formula to do that:

```
IF SUM([Number of Records]) <> SUM([FIXED Number of Records]) THEN "FILTER(S) IN
USE"
ELSE ""
END
```

Because there are only two outcomes, you can also write this formula as `IIF(SUM([Number of Records]) <> SUM([FIXED Number of Records]),FILTER(S) IN USE","")`.

The third step is to create a sheet that contains your Filter(s) in Use alert. To do so, create a new worksheet, and drag the newly created Filter(s) in Use Alert calculated field to the Text Marks Card. If the calculation is working, you will not see anything show up because there are no filters being applied to the sheet and the total number of records in the file matches the number of records on the sheet. You can format the alert to be the color, font, and size of your choosing after the next step.

Now, the key to the entire tip. To get the Filter(s) in Use Alert calculated field to properly calculate whether the filtered number of records matches the fixed number of records, ensure that any filter that you want to be included in the alert is applied to this worksheet.

To apply a filter to multiple sheets (including the Filter(s) in Use Alert sheet), go to any sheet containing the filter, and then, on the Filters Shelf, right-click the filter (or click the down-arrow that appears when you hover over the filter), hover over Apply to Worksheets and then click Selected Worksheets:

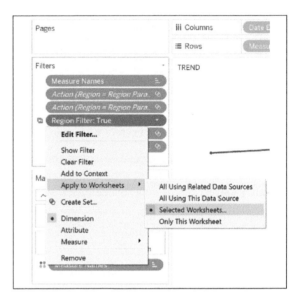

After you click Selected Worksheets, you will be presented with a list of all the worksheets in the workbook where you can choose to apply the filter to the newly created Filter(s) in Use Alert worksheet:

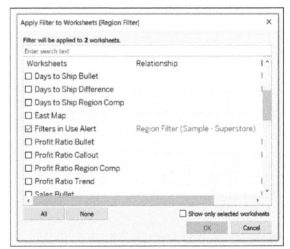

After I apply this filter to my Filter(s) in Use Alert sheet, my final view looks like this:

Note that by picking and choosing individual filters to apply to this sheet, you are setting the rules for how rigid the alert will be. In my example, changing the Date filters will *not* show an alert because I haven't applied the Date filters to the Filter(s) in Use Alert worksheet. Changing the Region filter *will* show an alert

because I have applied the Region filter to the Filter(s) in Use Alert worksheet. If you want the alert to show up if *any* filter that is applied, simply apply every filter being used in the workbook to your Filter(s) in Use Alert worksheet.

The last thing to do is add this sheet to your dashboard and soak in your newfound peace of mind!

CHAPTER 84

How to Add Custom Integrated Insights to a Dashboard

When I discuss data visualization strategy, I often talk in terms of explanatory (i.e., descriptive) and actionable (i.e., prescriptive) analytics. Explanatory views are the most common and can be thought of as high-level overviews of *what* happened. Most weekly or monthly reports, particularly if they're consumed as static documents, are explanatory dashboards. Explanatory dashboards certainly provide value and have a place, but they're only a starting point for deeper analysis.

Actionable views go a step further than explanatory views and attempt to answer *why* something happened and provide ideas for how to act. For example, an explanatory view might show that sales spiked 25% last month, and an actionable view would add value by showing it was the 15% off email campaign that caused the spike. Ideally, actionable views also *prescribe* what should be done about the insight. For example: we saw the email campaign worked; next we should test the discount amount to maximize our profit ratio.

Even though explanatory or descriptive insights can be easily ascertained by most audiences, actionable or prescriptive insights and recommendations are much more difficult to come by. I know some really smart people are working on ways to replace analysts with robots, but in the meantime, I found a simple way to allow you and your end users to add their own actionable insights and recommendations to a Tableau dashboard.

How to Add Actionable Insights and Recommendations with Parameters

To help illustrate, for this exercise, we use the *Super Sample Superstore* dashboard (*https://oreil.ly/6Unk0*). As you can see in the following screenshot, there are three different insights at the bottom of the Actionable view:

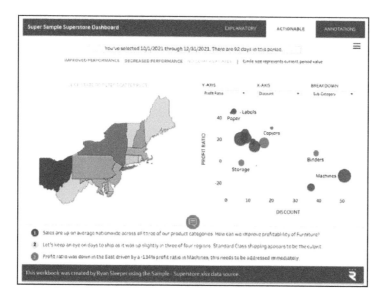

These three insights were entered on the Annotations tab:

The end user has the ability to type their insight into a free-form text box and choose whether the insight is positive, neutral, or negative. The rating is mapped to a color-coded indicator and both the indicator and insight show up on the Actionable view!

To accomplish this, we begin by creating a parameter with a data type of String. Here are the settings for the first parameter:

Note the allowable values were set to All to allow the end user to type any string.

Parameters do very little on their own, so to get the end user's input onto a view, I also created a calculated field that shows whatever is typed into the parameter's free-form text box:

Now this calculation can be placed on the Label or Text Marks Card of a sheet and formatted to your liking. Be sure to show the parameter control by right-clicking the parameter and then, on the menu that opens, choose Show Parameter Control. Now, whatever is typed into the parameter control will show up on the sheet.

For the colored circles, let's set up another string parameter with the options of Positive, Neutral, or Negative:

We're now ready to create the indicator sheet. On a new sheet, let's change the mark type to Circle and place the Insight 1 calculated field on the Label Marks Card:

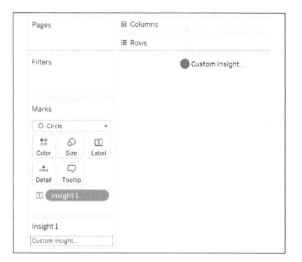

Now that there is a field on the Marks Shelf, I can add the Insight 1 Indicator parameter directly to the Color Marks Card. After showing the parameter control for the Insight 1 Indicator parameter, you can click through each of the three options and map the colors of your choice:

There are several options for how to align the insights with the circle indicators. You could keep the default as I've shown, but it would be tricky to align multiple insights. Option 2 would be to have the circles and text on separate sheets. Option 3 would be to add a placeholder field (i.e., MIN(0))

to the Columns Shelf and edit the axis to something like –5 to 100 to always have the circles left aligned.

Repeat these steps for the number of insights you want you and your end users to enter. The *Super Sample Superstore* dashboard allows up to three customized insights.

Now that you have provided the ability to seamlessly enter customized insights, users can save their own views, including insights on Tableau Server. For example, you can open a weekly report, add your insights, and then save and send it out. This approach even works on Tableau Public, and the insights persist when you use the share link!

CHAPTER 85

How to Make Automated Insights in Tableau

In Chapter 84, we examine how to add custom integrated insights to a Tableau dashboard; a feature that does not come out-of-the-box in the software. This is an important prescriptive tactic that helps explain your insights and actionable recommendations. *But what if you could automate those insights?* Well, you can compute just about any calculation in Tableau. The results of these calculations can then be combined with text to provide automatic insights to you and your audience.

In this chapter, we reverse engineer a Tableau Public visualization to show how automatic insights are created, I demonstrate how to concatenate text and computed string results to automatically spell out full sentences in Tableau, and we use LOD expressions to create an automatic insight that compares the performance of a specific dimension member to a benchmark.

An Example of Automatic Insights in Tableau

I haven't shared this before, but my favorite visualization I've ever personally created is *Your Salary vs. a MLB Player's Salary* (*https://oreil.ly/y3DsB*)—and it even has donut charts on it!

Questionable practices aside, there are several elements I like about this one:

- The user becomes part of the story by inputting their own salary.
- The bars appear to animate (an illusion created by ordering the bars so that they would render sequentially).
- Simple design.
- Cross-branding and linking.

There's even an Easter egg if you type in a salary greater than a million dollars!

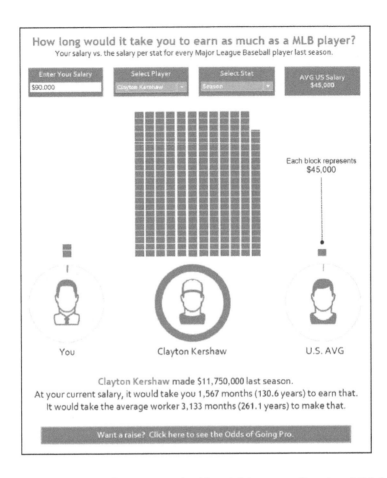

Picking your favorite visualization can be like picking your favorite child, but what puts this one over the top for me are the automatically generated insights at the bottom. Not only does the end user become part of the visual representation of the salary comparison, but they are reaffirmed as part of the story with a caption that becomes customized to their personal salary.

We've become desensitized to hearing the gigantic salaries of professional athletes these days, but when you see the visual and read the unique caption, it has an almost visceral effect. What's great about these features is that I only had to set them up one time; from then on, Tableau does the rest of the work for me to generate the captions automatically.

You can download any of my Tableau Public visualizations (*https://oreil.ly/XMjQt*) and then reverse engineer them to see exactly how they were created. Here's a look at the automated insight:

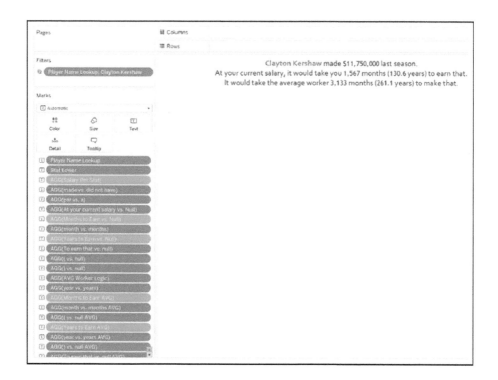

As you can see, there are quite a few fields on the Text Marks Card. In fact, here's a closer look that can be accessed by clicking on the Text Marks Card and then clicking the ellipsis next to the Text box:

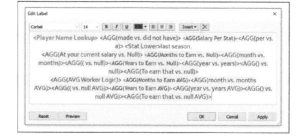

Note that there are only two words in this entire word processor that are hardcoded (which I know because hardcoded words don't have gray shading behind them): *last season*. Everything else is computed dynamically to spell out the appropriate sentence based on the user's salary, player, and statistic selections.

Admittedly, depending on the complexity of your insights, this can become quite a rabbit hole. In my use case, I discovered all kinds of scenarios that had to be accounted for in the underlying calculated field for each element of the sentences. For example, American League pitchers would not have hitting statistics; sometimes the result would be months instead of years; sometimes fields needed to be singular instead of plural; and more.

I recommend breaking each element of your insight into a separate calculated field for easier management.

For examples of how mine were created, simply download the workbook from Tableau Public (*https://oreil.ly/ y3DsB*), right-click any field that starts with an equal sign (=), and then click Edit:

```
Stat Lower                    Salary#csv (Salary.csv)                    ×

IF [Parameters].[Stat] = "RBI" THEN "RBI "
ELSEIF [Parameters].[Stat] = "Season" THEN ""
ELSE LOWER([Parameters].[Stat])+" "
END

                                                                    ▸


The calculation is valid.        9 Dependencies ▾    Apply      OK
```

```
IF [Parameters].[Stat] =
"RBI" THEN "RBI "
ELSEIF [Parameters].[Stat] = "Season"
THEN ""
ELSE LOWER([Parameters].[Stat])+ " "
END
```

The goal is to combine these individual elements that spell out full sentences with proper spelling, punctuation, and grammar.

How to Concatenate Strings or Text Fields in Tableau

The most flexible way to concatenate fields in Tableau to form full sentences is to put the fields on the Text Marks Card and then reorder the fields so that the sentences make sense. This approach allows you to alter the format of individual fields and use a combination of text and numbers. The one drawback to this approach is that it can be a lot to manage. The one paragraph of text in the *Your Salary vs. a MLB Player's Salary* visualization has *21* different fields on the Text Marks Card.

If you prefer to consolidate some fields to help make them easier to keep track of, you can also concatenate strings by creating a calculated field. The trick to concatenating strings in Tableau is to simply *add* the strings together within a calculated field. Here's a calculation that writes the first sentence in the visualization's caption:

```
First Sentence                Salary#csv (Salary.csv)                    ×

ATTR([Player Name Lookup])
+ ' '
+ [made vs. did not have]
+ ' $'
+ STR([Salary Per Stat])
+ ' '
+ [per vs. a]
+ ' '
+ [Stat Lower]
+ ' last season.'

                                                                    ▸


The calculation is valid.                          Apply      OK
```

```
ATTR([Player Name Lookup])
 + ' '
+ [made vs. did not have]
+ ' $'
```

```
+ STR([Salary Per Stat])
+ ' '
+ [per vs. a]
+ ' '
+ [Stat Lower]
+ ' last season.'
```

There are a few nuances to this calculation:

- You cannot mix aggregates and non-aggregates in a Tableau calculated field, so I had to add ATTR before the Player Name Lookup field (the only field that wasn't an aggregate).

- The lines that have ' ' are adding spaces between the elements in the sentence.

- You cannot add String and Float data types so I had to convert the one number in this calculation (Salary Per Stat) to a String by wrapping it in the STR function.

- When you convert a Float to a String, you lose the number formatting; this is why I added a dollar sign before the Salary Per Stat field.

- In string calculations, it doesn't matter whether you use single tick marks or double quotation marks.

You would then place this calculated field on the Text Marks Card just as I did in this example with the individual elements. The difference in this case is we've consolidated five fields into one to create the first sentence!

Using LODs to Create an Automatic Insight to Compare Against a Benchmark

So let's put some of these ideas to work to create a new use case. One of my favorite applications of Tableau LOD expressions is to create benchmarks. For illustration, suppose that our benchmark is the Machines sub-category in the Sample – Superstore dataset and we want to create an automatic insight to compare the performance of each sub-category to Machines.

First, we set up a calculation that isolates the performance of the Machines sub-category. Here's the formula:

```
{FIXED: SUM(IF [Sub-Category]
= "Machines"
THEN [Sales] END)}
```

This version of the benchmark calculation will always show us the SUM of sales for the Machines sub-category across the entire file. If you would

---

prefer your benchmark to be filtered by additional dimensions later in your analysis, I recommend using an EXCLUDE expression instead. The formula would be {EXCLUDE [Sub-Category]: SUM(IF [Sub-Category] = "Machines" THEN [Sales] END)}.

Let's imagine that we want our automatic insight to say "The [Sub-Category Selected] sub-category performed X% [Better or Worse] than the Machines benchmark."

There are four elements in this insight: hardcoded text, the sub-category selected, the percent compared to the benchmark, and the word *better/worse*. Two of these four elements need to be computed: the percent change and the word *better/worse*:

Here's the first formula:

```
SUM([Sales]) /
SUM([Benchmark:
Machines]) - 1
```

For best results, let's change the default formatting of this calculated field to percentage with no decimal places. In the Measures area of the Data pane, right-click the calculated field, hover over Default Properties, and then, on the menu that opens, choose Number Format.

Here's the second formula

```
IF [Sales vs. Machine Sales]
(from the first formula) > 0
THEN 'better' ELSE
'worse' END
```

You can also write this formula as IIF([Sales vs. Machine Sales] > 0,'better','worse').

Now that we have all the components, let's add each to the Text Marks Card of a new sheet:

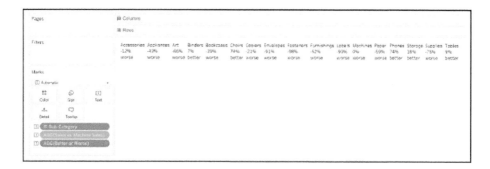

Just glancing across the table, I can see that the negative percentages have the word *worse* and the positive percentages have the word *better*, so it's passing the eye test so far. To do a full quality check, you could simply add the Sales measure to the Text Marks Card, as well, and make sure the percentage calculations are checking out.

This insight is meant to compare only one sub-category to Machines at a time, so let's add a single value drop-down filter and filter the view to the Accessories sub-category:

Lastly, click the Text Marks Card to add the hardcoded text and put the dynamic elements in the appropriate place within our automatic insight:

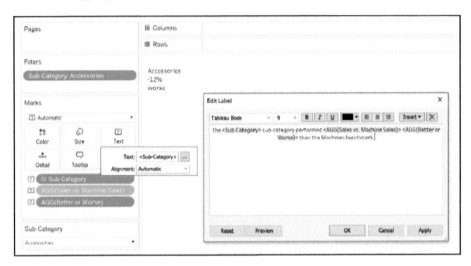

Now when I choose a different sub-category dimension member to focus on, the insight updates to automatically reflect the performance of that specific dimension member to our benchmark:

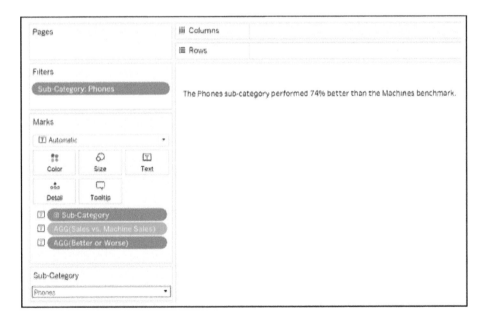

I can now put this sheet on a dashboard, and Tableau does the rest of the work for me regardless of the sub-category that the end user selects. You can use this approach to automatically compare against a competitor, goal, and/or performance last year.

CHAPTER 86

How to Make Color-Coded Navigation Using Dashboard Actions

As is often the case, new client requirements inspire new approaches that are made possible through Tableau's flexibility. In one such instance, a need came up to create a navigation that doubles as a color legend. You can set up color legends to highlight dimension members, but this navigation needed to *filter*. A filter alone wouldn't work because filters don't include color encoding.

So let's combine the two using a highlight table and dashboard actions! This approach provides several benefits including formatting flexibility, improved dashboard processing time, and—what this section is all about—a better user experience.

If you are familiar with this technique, you likely know that the Square mark type can limit your ability to make nice, evenly sized rectangles that fill all the available space in each cell of the highlight table. This tutorial shows you a trick for getting around that, too, producing perfect rectangles every time!

How to Convert a Highlight Table into a Navigation + Color Legend

To illustrate how to combine a navigation and color legend using a highlight table with dashboard actions, we re-create the following small dashboard. This dashboard uses a Tableau dual-axis waterfall chart (Chapter 37) to show how profit is trending in the Sample – Superstore dataset:

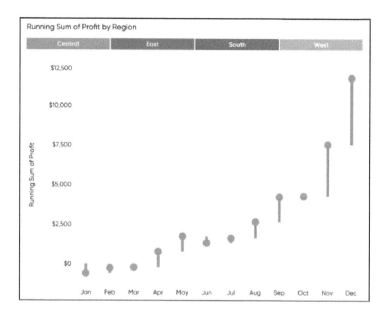

Clicking a region tab at the top of the window filters the chart:

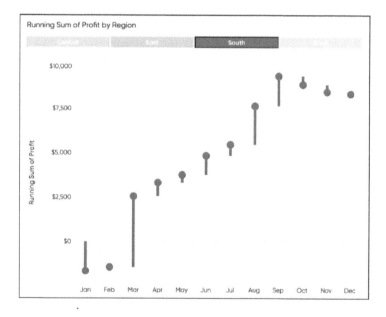

The tab bar at the top is created by setting up a control sheet that includes the dimension members you want to make available to end users. To set up the view, begin by creating a new worksheet, place the dimension of interest on the Columns Shelf, change the mark type to Square, and then place the dimension of interest on the Color Marks Card as well as the Label Marks Card:

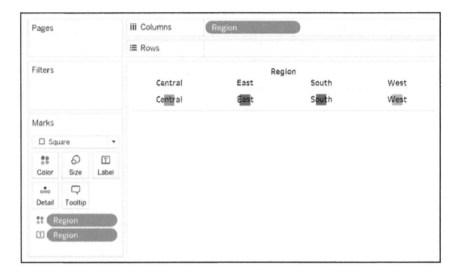

This is a good start, but obviously we need to make the squares larger; ideally filling all the white space between each button. This is where most folks run into some formatting issues. Here's how the view looks after clicking the Size Marks Card to increase the size of the square marks and dragging the slider all the way to the right (making the marks as large as possible):

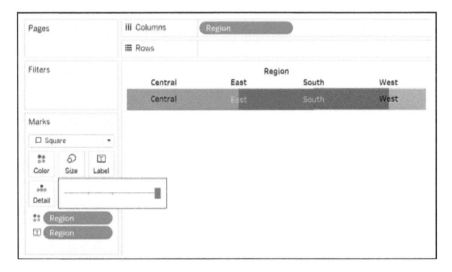

Yikes—not very effective! The orange square that represents the Central region is laying on top of the others so we do not see as much of the red, blue, or turquoise squares. Luckily, there is a simple solution. Drag the dimension of interest to the Columns Shelf a second time:

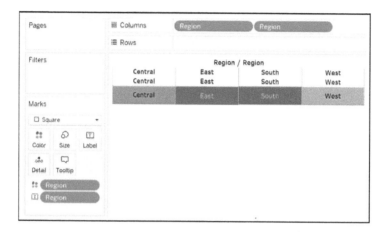

There are a few more things I like to do to a view like this before adding it to a dashboard:

- Hide the headers by right-clicking them and then deselecting Show Header
- Format the column and row separators for the table by right-clicking anywhere on the table and then choosing Format
- Ensure that the labels are always a consistent color by clicking the Label Marks Card and changing the font format

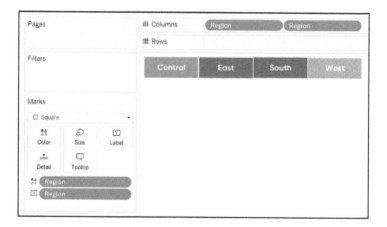

The color legend/tab bar is now ready to be added to a dashboard. The user experience of having the dashboard filter to the dimension member clicked from the table is accomplished through dashboard actions. Simply add a dashboard action that suits your needs; here is a look at the dashboard action for the use case illustrated in this chapter:

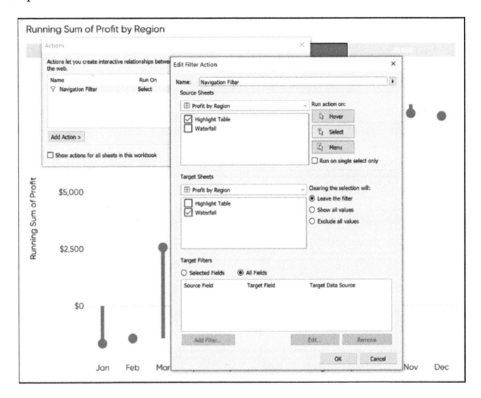

Now, clicking any region in the highlight table will filter the Waterfall view. Because this particular chart can be confusing to interpret when all four regions are being shown together, I also changed the settings in the "Clearing the selection will" section to "Leave the filter." This means that if the selection is cleared by pressing the Escape key, the selection will persist until another region is clicked.

CHAPTER 87

How to Make a Parameterized Scatter Plot

One of my favorite actionable or prescriptive tactics is to provide a scatter plot that the user can build themselves—even if they don't know how to use Tableau! The *parameterized scatter plot* is very actionable, not only because the audience literally acts by creating its own analysis, but it often reveals new correlations and outliers that the business can act upon.

Scatter plots have several advantages: they're able to show many data points at once, help illustrate correlations, and create a natural four-quadrant segmentation.

This chapter shows you how to make scatter plots even better by allowing your end user to choose the measure displayed on the y-axis, the measure displayed on the x-axis, and dimensional breakdown of the marks on the view.

How to Let Your Audience Build Their Own Scatter Plot in Tableau

For this tutorial, we rebuild the scatter plot from the Actionable dashboard in the *Super Sample Superstore* workbook (*https://oreil.ly/6Unk0*). In the scatter plot on the right of the dashboard, the user can choose the measure on the y-axis, the measure on the x-axis, and the dimensional breakdown of the circle marks:

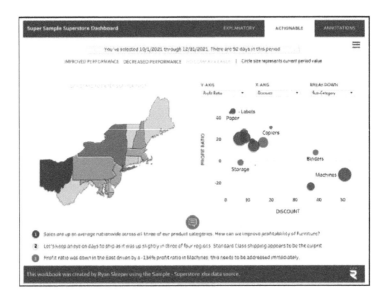

To allow an end user to build their own scatter plot, we use the tactic outlined in Chapter 64 in *Practical Tableau* (O'Reilly, 2018).

First, you must set up a parameter with a data type of String for each of the three choices (y-axis, x-axis, and dimension breakdown). The allowable values for each will be the choices that the end user can eventually access to build their scatter plot. Here's how each of the three parameters look:

Y-Axis

X-Axis

Scatter Plot Detail (the dimensional breakdown)

Parameters do almost nothing on their own, but I will share one useful application at the end of this chapter. Most often, parameters need to be used within a calculated field to give Tableau instructions for what to do with each allowable value. Here are the calculations for each of three parameters, which use CASE/WHEN logic to display the appropriate measure or dimension when they are selected from our list of allowable values in the parameters.

Note that when you see the measure multiplied by 100, I'm simply making ratios more readable by moving the decimal two places to the right.

Y-Axis:

```
CASE [Parameters].[Y-Axis]
WHEN "Sales" THEN [CP Sales]
WHEN "Profit" THEN [CP Profit]
WHEN "Quantity" THEN [CP Quantity]
WHEN "Discount" THEN [CP Discount]*100
WHEN "Returns" THEN [CP Returns]
WHEN "Days to Ship" THEN [CP Days To Ship]
WHEN "Profit Ratio" THEN [CP Profit Ratio]*100
END
```

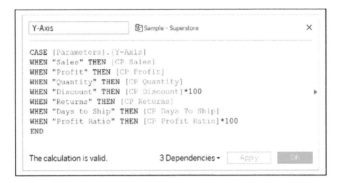

X-Axis

```
CASE [Parameters].[X-Axis]
WHEN "Sales" THEN [CP Sales]
WHEN "Profit" THEN [CP Profit]
WHEN "Quantity" THEN [CP Quantity]
WHEN "Discount" THEN [CP Discount]*100
WHEN "Returns" THEN [CP Returns]
WHEN "Days to Ship" THEN [CP Days To Ship]
WHEN "Profit Ratio" THEN [CP Profit Ratio]*100
END
```

Scatter Plot Breakdown

```
CASE [Scatter Plot Detail]
WHEN "Segment" THEN [Segment]
WHEN "Ship Mode" THEN [Ship Mode]
WHEN "Customer Name" THEN [Customer Name]
WHEN "Category" THEN [Category]
WHEN "Sub-Category" THEN [Sub-Category]
WHEN "Product Name" THEN [Product Name]
END
```

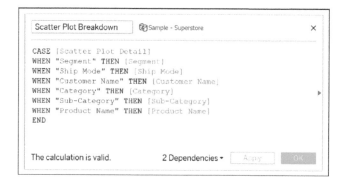

We now have all the elements needed and can build the scatter plot! To do so, put the Y-Axis calculated field on the Rows Shelf, the X-Axis calculated field on the Columns Shelf, and the Scatter Plot Breakdown calculated field on the Detail Marks Card. Let's also change the mark type to Circle and increase the size of the marks using the Size Marks Card so that we can see them better:

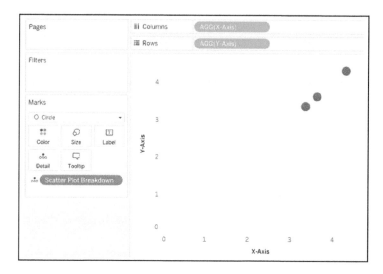

Everything is working so far, but the current value of both the Y-Axis and X-Axis parameters is Days to Ship, so we're looking at the same metric on both axes broken down by Segment (the current value of the Scatter Plot Detail parameter). To allow our end users to build their own scatter plot and choose the measures and dimensions that are relevant to them, we must right-click each of the three parameters and then, on the menu that opens, select Show Parameter Control. This displays three drop-down menus on which the end user can now select from the full list of allowable values coded in the parameters.

Here's how the view looks after we show the parameter controls and choosing Sales for the y-axis, Discount for the x-axis, and Sub-Category for the dimensional breakdown:

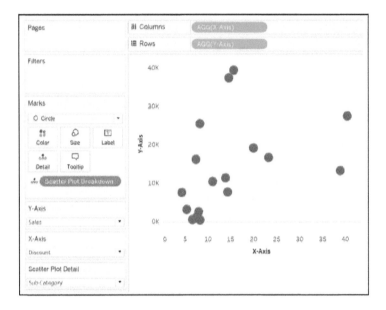

The next step is optional, but in the final version of my parameterized scatter plot, I colored the circles by period-over-period change and sized the circles by a fourth parameter selection.

Remember when I mentioned earlier that parameters do almost nothing on their own? One thing they can do is provide axis labels by placing them on the appropriate shelf. For example, if we place the Y-Axis parameter on the Rows Shelf and the X-Axis parameter on the Columns Shelf, each axis label will display the name selected in the respective parameter:

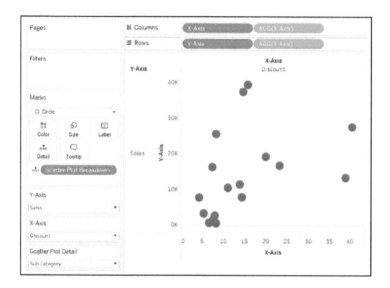

Of course, we can further format this by rotating the labels, hiding the words *X-Axis*, and so on.

Scatter plots were already a favorite of mine due to the benefits outlined in Chapter 22, and now end users can take advantage of these benefits by building them on their own. They don't need to know how to author in Tableau; they simply need to know how to make their selections using parameter controls.

CHAPTER 88

How to Make a Global Filters Tab

I often see corporate Tableau *dashboards* built with nothing more than a large text table and a dozen or more filters along the top or side of the crosstab. Sadly, I'd even go as far as to say that this is what *most* dashboards look like when I enter a new engagement. Although there is some value in using Tableau purely as a querying tool —most notably the way this type of view helps validate raw data—it misses the point of how data visualization can help us analyze data.

So how do we get users to evolve past the three-decade-old, Excel-like views? If I have stakeholders who simply can't let go of filters number seven and above, I like to build in a dashboard element that I call the *global filters tab*. This feature helps prioritize filters, but also provides the flexibility to access lesser-used filters. This chapter shows you how to create a global filters tab in Tableau so that you can keep your dashboards clean while keeping your stakeholders happy.

How to Clean Up Dashboards by Moving Filters to Their Own Worksheet

I'm a minimalist when it comes to dashboard design. I work hard to maintain focus and prioritize only the most pertinent elements for each individual view. For this reason, I actually don't have a public corporate example of a global filters tab to share with you. So for this exercise, we add one to the Explanatory dashboard in the *Super Sample Superstore* workbook (*https://oreil.ly/6Unk0*):

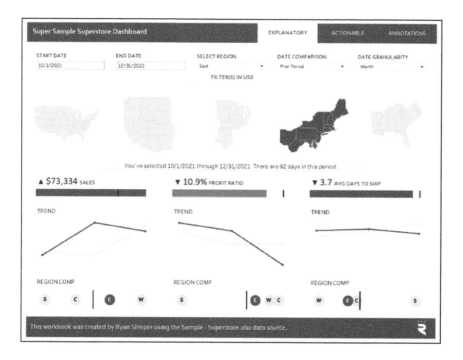

As you can see on the filters row, there are five parameter controls being used to filter the view. Let's assume that in addition to filtering the date range, region, and comparison period, our stakeholder requested that we add 10 more filters: Customer Name, Segment, Ship Date, Country, State, Returned, Ship Mode, Category, Sub-Category, and Product Name.

I hope you laughed at that because this does happen. I don't know about you, but I'm not muddying up this beautiful dashboard by forcing 10 more filters onto it. Instead, let's create a global filters tab; this acts as a proverbial rug to sweep all our dirt under.

To create a global filters tab, start a new worksheet and put anything on the Text Marks Card. For such occasions when the dimension is completely arbitrary, I like to create a calculated field called *Blank* that contains just two quotation marks:

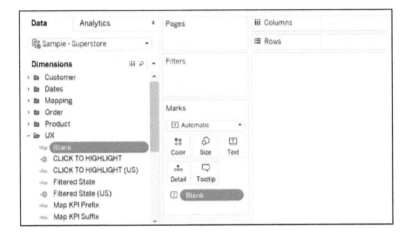

You can also add the two quotation marks in the flow of the analysis by double-clicking in the blank space below the Marks Cards on the Marks Shelf, but you would need to remember to then drag the newly created field to the Text Marks Card.

After you have something on the Text Marks Card, right-click each field by which you would like to filter and then, on the menu that opens, choose Show Filter:

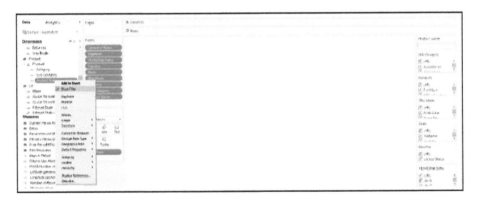

By default, all of the filters will be shown along the right side of the worksheet in their default format. You can reorder these filters, place them in different locations on the worksheet, and/or change their format to your liking:

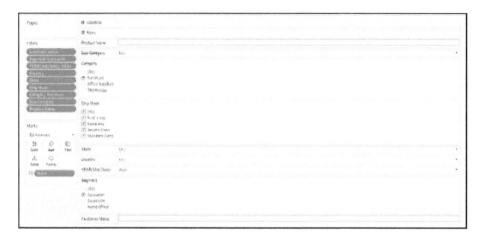

Next, ensure that all of the filters are global by right-clicking each one on the Filters Shelf, hovering over Apply to Worksheets, and then choosing All Using Related Data Sources (or All Using This Data Source):

Now when any filter is changed on this tab, it will filter every individual worksheet in this workbook. You can help your users navigate to this worksheet by adding a Button object wherever you would like on the dashboard and link it to the global filters tab:

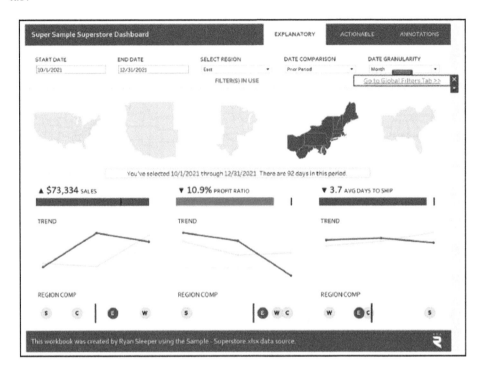

Now, when a user clicks the Go to Global Filters Tab button, they are redirected to the Global Filters tab and can update their global filters. If this tactic is particularly relevant to you, you can always take my Global Filters worksheet a step further and make it into a nicely formatted dashboard of its own. That option comes with the benefit of allowing you to add a back button that brings the user back to the original dashboard.

CHAPTER 89

How to Make Indicators with Custom Shape Palettes

Like the stock ticker gauge described in Chapter 43, most shapes are not the best choice for visualizing *magnitude*, but just like sometimes when you need something other than a bar chart, every now and then you need a gauge other than a bullet graph. I like the custom shape indicators I'm sharing here because they immediately and intuitively convey performance, have a minimalist design, and provide a starting point that can be linked to deeper analysis.

This chapter shows you how to find and format custom shapes for your dashboard, install those shapes in your Tableau Repository so you can map them to your data, write a calculation to dynamically change the shape being displayed based on performance, and use dashboard actions and/or the Viz in Tooltip feature to link to deeper analysis.

How to Create an Engaging, Explanatory View by Using Custom Shapes

By the end of this chapter, you will be able to make indicators using shapes of your choice that change based on performance.

In my example, the indicators turn to a green smiley face if the month-over-month change for a KPI was positive, a yellow neutral face if the month-over-month change was negative but within 10%, and a red frowning face if the month-over-month change was worse than 10%. Of course, you can use any images and performance thresholds that you want.

The first thing we need are the images that indicate the change in performance. Here are a few ideas for obtaining images:

- Make them. I use Adobe Photoshop and Illustrator, but often Microsoft Power-Point will suffice.

- Buy them. The images in my example are a stock Illustrator file I purchased from iStock (*http://istockphoto.com*) for $12. If that's too rich for your blood, I also like the selection of icons available at The Noun Project (*http://thenounproject.com*), where you can download as many as you want for $39.99 per year. There are countless others, but these are the two I primarily use.

- Search them. To keep your positive karma in the universe, I recommend using Google Image Search: in the lower-right corner of the screen, click Settings > Advanced search. At the bottom of advanced settings is a place to choose the appropriate usage rights for your purposes.

For best results when using images in Tableau, in general, create the image or search for images in a PNG format. You can create and save PNG images with transparent backgrounds, allowing you to use the image in Tableau on different background colors or to reveal underlying marks (like you might need on a scatter plot).

Here is one of my indicators in Photoshop before I saved the image as a PNG file; the checkered background indicates the portion of the image that will be transparent:

After you have all the images you want to use in your visualization, you must place the files in the *Shapes* folder within the Tableau Repository on your computer. You can find this at Documents > My Tableau Repository > Shapes. Each folder within the *Shapes* folder corresponds with a shapes palette within Tableau Desktop that you can use to map shapes to dimension members.

You can copy and paste the image files into a *Shapes* folder or create your own. I like to create new folders, which will generate a separate palette that I can eventually access when I'm using Shape as my mark type in Tableau. Here's how my folder for *Indicators* looks on a PC.

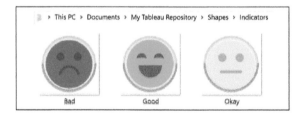

We're now ready to open Tableau and build the indicator gauges. To create gauges that represent month-over-month performance or performance compared to a goal, you need two calculated fields for each measure:

- The first to compute the comparison
- The second to compute the Good/Okay/Bad thresholds (or whatever you are using)

For ease of illustration, let's pretend that we're comparing entire months to one another across all four years of data in the Sample – Superstore dataset, and that we want to compute the percent change between the month of March and February. You will need to adjust this calculation to include your two comparison points, but here is the formula for percent change:

```
([Current Performance] / [Comparison Performance])-1
```

If we're comparing March to February, the formula is as follows:

```
SUM(
IF MONTH([Order Date]) = 3
THEN [Sales] END)
/
SUM(
IF MONTH([Order Date]) = 2
THEN [Sales] END)
-1
```

For the second calculated field, let's use IF/THEN logic to create three different scores that will eventually be mapped to my custom shapes.

Here's how they'll map: if the current performance outperformed the comparison, we'll call that *Good*; underperformance within 10% will be described as *Okay*, and underperformance of greater than 10% is *Bad*.

Here's the formula:

```
IF [Month over Month Sales] > 0 THEN "Good"
ELSEIF [Month over Month Sales] > -.1 THEN "Okay"
ELSE "Bad"
END
```

To make the gauge, start a new worksheet, change the mark type to Shape, and put the calculated field that is computing the performance thresholds onto the Shape Marks Card:

In this example, March has outperformed February, so the indicator has been classified as Good. So far so good, but now we need to map our custom shapes to each of the three performance scores.

The first one is easy because it is already on the view. Just click the Shape Marks Card to access your available shape palettes and then use the Select Shape Palette drop-down box to find the custom shape palette installed in your Tableau Repository:

If you installed the custom palette while Tableau was open, you will need to click the Reload Shapes button in the lower-right corner of this dialog box before the new palette will appear.

To map a performance score to a specific shape, ensure that the dimension member is selected from within this dialog box and then click the shape that you want assigned to the dimension member:

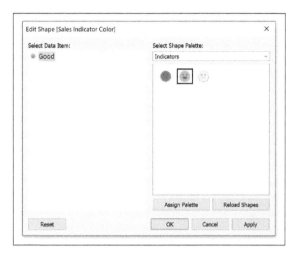

Click OK; you now have your first indicator mapped. We'll see a green smiley face any time the current performance outperformed the comparison:

We have only mapped one of the three outcomes, so should the performance change to Okay or Bad in the future, at this point we would see only default shapes. How, then, do you map the other two outcomes if they're not on the view? You could manipulate the dates in the month-over-month calculated field to

attempt to find months that match the other scoring thresholds, but I have an easier way.

To map discrete scoring scenarios (like we have here) that are not on the view or have yet to appear in your data, temporarily hardcode numbers that match the unseen scenarios in the underlying calculated field. I like to do this by using the aggregation MIN and typing in numbers that match the additional outcomes. For example, to create the Okay scenario, I edit the calculated field that is computing the comparison and overwrite what is there with `MIN(-.01)`. Just make sure you copy and paste the original formula back into the calculated field after you have mapped all the outcomes.

-.01 is an underperformance of 1%, which in this example, results in a scoring outcome of Okay. We can now map the neutral yellow face to this scenario:

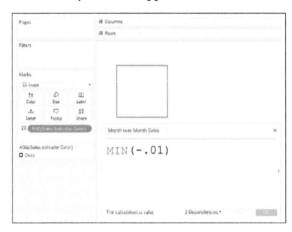

Repeat this step for as many outcomes as you have and then paste your original formula back into the calculated field. The mapping now works for every scenario, even if it hasn't appeared in your data yet!

Add Value with Tableau's Viz in Tooltip Feature and/or Dashboard Actions

This type of gauge is extremely descriptive in nature, providing only a very high-level overview on which areas of the business are healthy or need attention. One way to add value to this custom shape gauge is to use Tableau's Viz in Tooltip feature, which we explore in Chapter 9.

To display a trend when the end user hovers over one of my custom indicators, let's first make a trend that looks at sales by Month of Order Date:

We can add this graph to the tooltip of the Sales indicator by navigating back to the sheet for the indicator, clicking the Tooltip Marks Card, clicking the Insert button in the upper-right corner of the Tooltip interface, and then hovering over Sheets:

After you choose the Sales Trend sheet, it will appear in your tooltip. You might need to adjust the height and width defaults to ensure that the graph is not cut off when being viewed within the tooltip. Here are my settings after changing the height and width to 500 pixels each:

After clicking the OK button, the user will see the 12-month trend when they hover over the indicator!

Of course, you can customize this Viz in Tooltip with whatever is relevant for your business and add fields to the tooltip such as current performance and comparison performance.

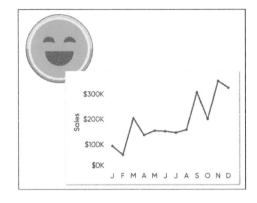

Another idea for adding value to these descriptive indicators is to use dashboard actions to link to deeper analyses. To do so, add the indicator to a dashboard and navigate to Dashboard > Actions in the top navigation. To navigate the user to a sheet within the same workbook, add a Go to Sheet action; to navigate the user to a view that lives on Tableau Server, Tableau Online, or Tableau Public, add a Go to URL action. This is a great technique for keeping your workbooks and dashboards uncluttered and only linking your audience to deeper views when relevant.

CHAPTER 90

How to Make a Timeline When Events Overlap

In Chapter 33, I share how to make a timeline in Tableau. This is one of my favorite techniques for creating a calendar and providing context to my end users. You can use timelines to show your audience when notable events occurred in the business that might be affecting performance. But what happens when there is more than one event on the same day? By default, the marks will overlap, preventing your audience from seeing the entire picture.

This chapter shows you how to make what I call a *jittered timeline* in Tableau. When events overlap on the same day, the calculated field shared in this tutorial automatically provides perfect spacing between the marks. This is also helpful if you ever want to control the alignment of marks on a jitter plot. The controlled spacing serves not only a practical purpose, but it also looks great!

How to Make a Jittered Timeline in Tableau

By the end of this chapter, you will be able to create a timeline in Tableau that automatically spaces marks when more than one event happened on the same day. You will be able to control the intensity of the jitter and position the marks to align perfectly on a zero baseline.

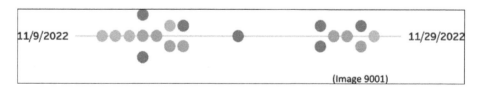

(Image 9001)

This is a technique I originally picked up from my friend, Lindsey Poulter, in her visualization, *13 Straight*:

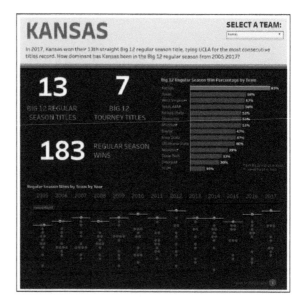

For the following illustration, let's pretend that we represent a company that will be doing a big marketing push the week of a Tableau Conference and Black Friday/Cyber Monday later in November. I have this schedule planned, but we want to visualize it as a timeline alongside other charts in Tableau:

To start the timeline in Tableau, let's place the Date field on the Columns Shelf as a continuous field at the *day* date part. To ensure that all the events in the dataset are represented on the view, let's place the Event dimension on the Detail Marks Card. Lastly, we need to change the mark type to Circle just so the events are easier to see:

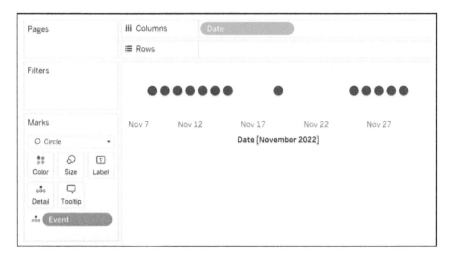

Here's the issue. Several of these dates have multiple events, and even though the visualization's level of detail is currently Event, we're not able to see that level of granularity. For example, November 12 has three events, but the three circles overlap. This gives the appearance to the audience that there is only one event that day. To visually separate marks on a timeline when more than one event occurs on the same day, we use a similar tactic that is used to create jitter plots.

Related: *Practical Tableau*, Chapter 46, "How and Why to Make Customizable Jitter Plots" (O'Reilly, 2018)

We will need to take the following concept a step further with the calculated field shared later in this chapter, but to illustrate why it's needed, let's start with a traditional jitter plot. Because there is a maximum of three events on any one day in this particular dataset, we place a jitter with an intensity of three rows onto the Rows Shelf. The formula is INDEX()%3:

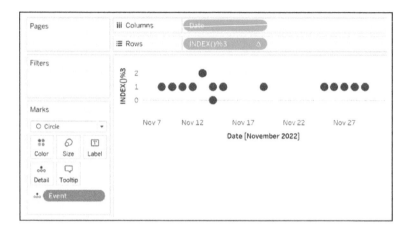

We also need to change the addressing of the INDEX() table calculation to be computed on Event instead of the default, Table (Across). To do this, click into a measure that has a delta symbol on it, hover over Compute Using, and then change the addressing:

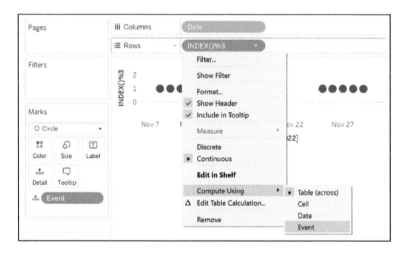

This assigns a number from 0 to 2 to each event and gives them unique locations on the y-axis. At this point, we've accomplished the objective of separating marks when there is more than one event on the same day. However, notice that whenever there are two events on one day, the events start at the value of 1 on the y-axis, and the two circles are not aligned perfectly with the days containing one or three events:

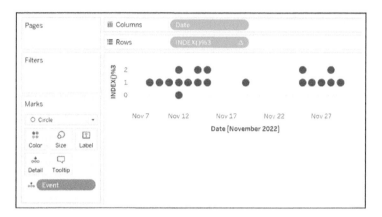

To make the circles line up perfectly, we need to *control the jitter* with the following calculated field:

```
INDEX()%
MIN({FIXED [Date]:
COUNTD([Event])})+
(IF MIN({FIXED [Date]:
COUNTD([Event])})=2 then -.5
ELSEIF MIN({FIXED [Date]:
COUNTD([Event])})=3 then -1
ELSE 0
END)
```

This uses the same foundation of a jitter plot where the INDEX() function is followed by a percentage sign and the intensity of the jitter. However, this time we are controlling the exact index value, or row number, for each scenario in our data.

The INDEX() function always starts at 0 and increases in multiples of 1, so our formula is saying when there are two events per date, start the values at 0 + −.5. This results in our first event being placed at −.5 on the y-axis and our second event being placed at .5 on the y-axis.

When there are three events on one date, the index values start at −1. This results in the first date being shown at −1 on the y-axis, the second value being shown at 0, and the third being shown at 1.

In this case, we use the ELSE statement to capture everything else, which in our sample dataset would be one event per date. When that is the case, the single events line up at 0. Here's how the timeline looks after we replace the original jitter with the newly created Controlled Jitter calculated field and, once again, we change the addressing to Event:

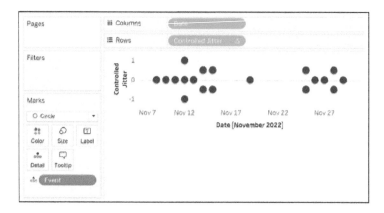

Note how the dates are not only separated but are aligned perfectly on a baseline of 0. Here are few more notes on ways in which you can use this:

- This works if your dataset has more than three events per day. You would just add scenarios to the calculated field. Just know that whatever number you type after each THEN statement will be the value that the events start on the y-axis.

- An alternative option to the approach I've shared here is to convert the timeline with overlapping events into a unit histogram by turning on stacked marks, as described in Chapter 36.

- This works even if you're not making a timeline view. You can make a *controlled jitter plot* when wanting to separate overlapping marks for any dimension. Just replace the Date field in the calculation in this exercise with whatever dimension has overlapping data.

- If you add levels of detail to the view, you will need to update the addressing accordingly to get things to line up. For example, here's how my view looks after coloring the circles by the Campaign Type dimension:

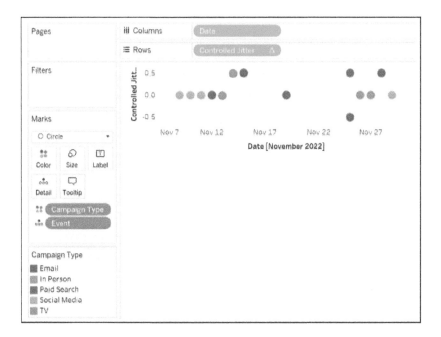

We've lost the nice alignment on the zero baseline. This one is not quite as easy as changing the addressing using the Compute Using option because we now need to use two different dimensions: Event and Campaign Type. To update the addressing so that both dimensions are included, you must right-click the Controlled Jitter calculated field and then choose Edit Table Calculation:

After we include both Event and Campaign Type in the specific dimensions being addressed, things are looking good again:

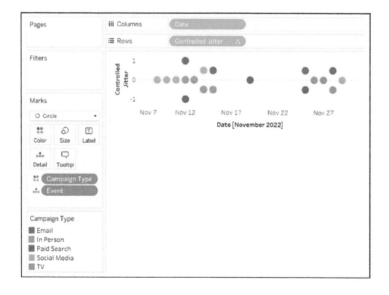

How to Make an Axis Range for a Tableau Timeline

One fun design idea is to show only the start and end of a timeline range. You can easily accomplish this by creating a sheet for the minimum date in your dataset, a sheet for the maximum date in your dataset, and then placing these two sheets along with your timeline on a dashboard.

To create the sheet for the minimum date, simply place the date field used to create your timeline on the Text Marks Card with an aggregation of MIN:

Then, do the same thing for the end date by starting a new sheet and placing the date field on the Text Marks Card with an aggregation of MAX:

Here is how the final view looks after we hide the original headers on the timeline view, updating the formatting to remove grid lines and axis rulers, and putting all three sheets next to one another on a dashboard:

For a live example of a controlled jitter plot/timeline built to show events when they are overlapping, see my visualization, *BLOCKBUSTER* (*https://oreil.ly/WkMQS*), in Chapter 95.

CHAPTER 91

How to Change Sort Order with Buttons in Tableau

Sorting marks within bar charts in ascending or descending order is good practice because it allows you to compare not only the preattentive attribute of height (or length), but also rank. By default, Tableau allows an end user to sort bar charts in ascending, descending, or data source order, but the functionality has several limitations, such as:

- Its appearance is very subtle and often goes overlooked.
- There are no formatting options.
- The sort does not work across multiple sheets.

The following technique overcomes all three of these limitations. This chapter shows you how to change the sort order of multiple sheets at the same time with the click of a button!

How to Simultaneously Change the Sort Order Across Multiple Sheets

By the end of this chapter, you will be able to toggle the sort between descending and ascending by clicking an up or down triangle, respectively:

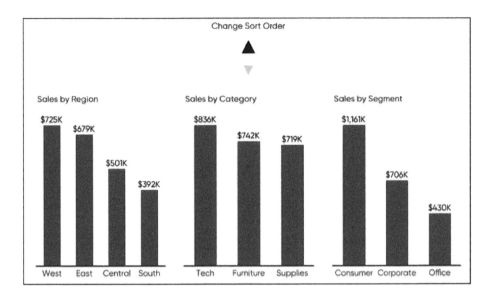

Step 1

Create a parameter with a data type of Integer and two values; one for descending and one for ascending. The values need to be 1 and –1, but you can set the Display As to whatever you want:

Step 2

Create a calculated field that equals the value 1. Whenever I create a calculated field containing one value, I like to add the aggregation of MIN to ensure that it's never aggregated in a different way:

Step 3

Create a calculated field that equals the value –1:

Step 4

Create a control sheet that will be used to select the sort order. The easiest way to do this is to create a new worksheet, change the mark type to Shape, place the Measure Names dimension on the Rows Shelf and Shape Marks Card, the Measure Values measure on the Detail Marks Card, and filter out every measure name except for the descending and ascending calculations from steps two and three. Here, I've also clicked on the Shape marks card to map a solid up triangle for descending and a solid down triangle for ascending:

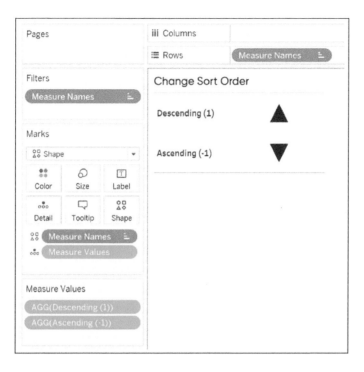

Step 5

Create a calculated field that multiplies the measure you are sorting by times the value in the parameter from step one. When Descending is selected, this results in the measure being multiplied by itself (so no change). When Ascending is selected, this results in the measure becoming negative. For the Sales measure in the Sample – Superstore dataset, the formula would be as follows:

```
SUM([SALES]) * [Sort Order]
```

Step 6

For each chart containing the measure you are sorting by, right-click the dimension on the Columns Shelf or Rows Shelf (depending on whether your chart is in a horizontal or vertical orientation), and then, on the menu that opens, click Sort to open the sort options, and sort the dimension in descending order by the newly created calculated field from step 5.

For this illustration, let's change the sort order across three separate sheets at the same time. The first is looking at Sales by Region, so we sort on the Region dimension:

Apply this sort order to the Sales by Category and Sales by Segment worksheets.

Step 7

Add the control sheet and all of the sheets that you want to sort to a new dashboard. I also hid the header names of the control sheet (Descending [1] and Ascending [-1]) by right-clicking the header and then deselecting Show Header:

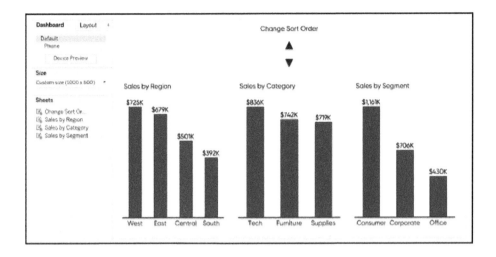

Step 8

On the menu bar at the top of the window, click Dashboard > Actions and then add a Change Parameter dashboard action. In the Actions dialog box that opens, click the Add Action button and then choose Change Parameter. In this example, we're setting this up so that if you click the Change Sort Order sheet, the underlying value will overwrite the Sort Order parameter with the measure value that is clicked:

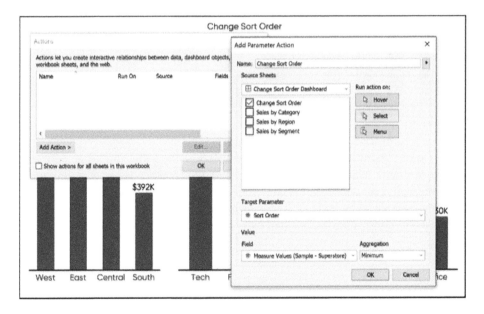

Remember, those triangles are mapped to the values 1 and –1. So, clicking on the up triangle overwrites the value in the parameter with the value of 1. This value of 1 then feeds the calculated field that is multiplying this value by the measure to control the sort order. Because the values are already being sorted in descending order and are now being multiplied by 1, you wouldn't see any change.

Clicking the down triangle overwrites the parameter value with –1. When Sales is multiplied by –1, it flips the best performer to the worst performer. Because the charts are still being sorted in descending order, the previously worst performers are moved to the front—we've created the illusion of an ascending sort!

The best part about this is we can use parameters across sheets and even data sources, so this technique will control the sort order across every sheet to which it is being applied:

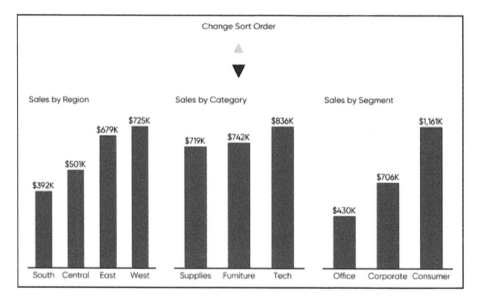

For ease of illustration, I sorted multiple sheets at the same time, but they all contained the same measure. This technique also works when using a variety of measures, but you must create the calculated field from step 5 for each measure being used and sort each chart by its respective sort order calculated field (i.e., Sales is sorted by Sales Sort Order, and Profit is sorted by Profit Sort Order). Every other step including the single parameter and single dashboard action are still shared.

CHAPTER 92

3 Innovative Ways to Use Transparent Sheets: Tip 1

As of Tableau Desktop 2018.3, authors can make worksheet backgrounds transparent. This means that you can float multiple worksheets on top of one another and still be able to see each individual layer. Transparent worksheets have unlocked new design possibilities such as floating all worksheets over a custom background image, but they also allow us to improve comparisons and context.

The next three chapters share three practical applications of transparent sheets to help improve your data visualization. You'll see how to add a trend behind a callout number, a timeline of key events to a line graph, and a custom segmentation to a scatter plot.

By the end of this series, you will know how to make all of the elements in this dashboard:

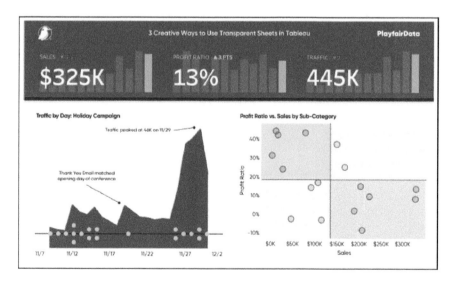

Add Context to Callout Numbers

Callout numbers, which are simply key information set with a large font size, are one of my favorite starting points for an analysis. I'll often list three or four of these along the top or left side of my dashboard to provide a 10,000-foot view of what's happening in the business. My audience can then decide whether they want to dig deeper on any of the metrics to find something more actionable.

Transparent sheets allow us to make callout numbers even better by providing a trend in the background.

Suppose that we have this callout number showing the current month's sales value and the month-over-month percent change:

Before we could create transparent worksheets in Tableau, I would put this on a dashboard by itself. Although the month-over-month percent change is helpful, it would be even nicer to see a trend for the last *twelve* months. Let's start a second worksheet that looks at sales by month:

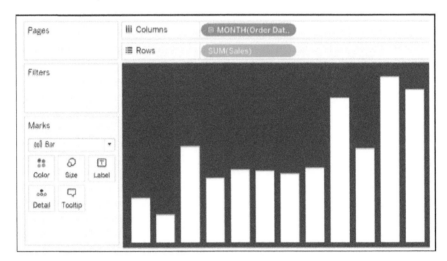

Since this will eventually be behind the Sales callout number, let's make the bars just 15% opaque by clicking the Color Marks Card and dragging the opacity slider left:

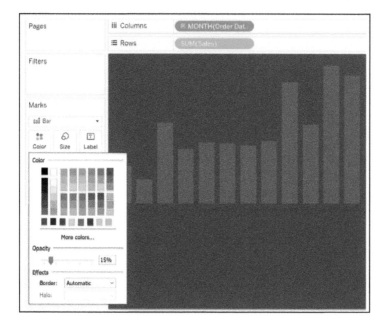

When we changed the opacity, all of the bars were updated to the same transparency. It would be a nice touch if we could make the most recent bar highlighted—I have a trick for that too!

Create a calculated field to show only the last value on the chart. The formula for Sales would be as follows:

```
IF LAST() = 0
THEN SUM([Sales])
END
```

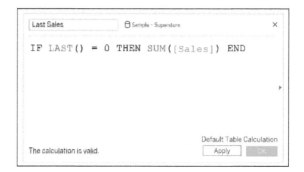

This is a table calculation that says if the difference between the current row and last row in the partition is zero, show the sales value; everything else will be null.

Now that we have the calculated field, I will create a dual-axis combination chart by placing the Last Sales calculated field on the right axis. When there are two measures on the Rows or Columns Shelf, each measure gets its own Marks Shelf that we can independently edit. One application of this is that I can make the primary bar chart 15% opaque and the last value 50% opaque, which creates a highlight effect:

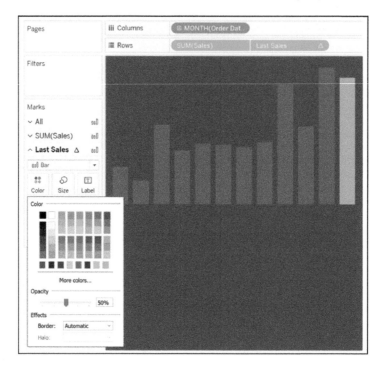

Now, we're ready to put both sheets onto a dashboard. Let's put the bottom layer on the *canvas* first:

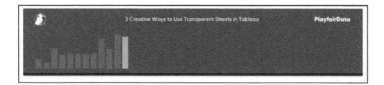

Float the callout number directly over the first layer with the same exact dimensions. You can change the dimensions and X-Y coordinates in the Layout pane of the Dashboard interface:

Here's the issue. When the background of the callout number worksheet is opaque, we cannot see the underlying trend below. As mentioned in the introduction, we can now make the background of the top layer transparent by right-clicking anywhere within the worksheet and then, on the menu that opens, choose Format:

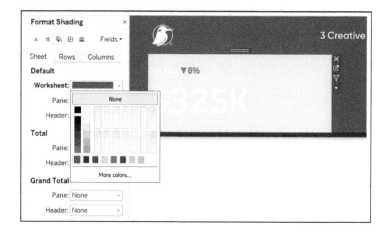

After changing the shading of the Sales callout number to None, we've created a nicely designed and practical element at the top of the dashboard!

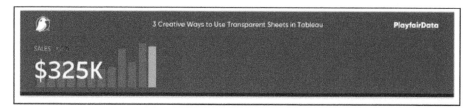

CHAPTER 93

3 Innovative Ways to Use Transparent Sheets: Tip 2

Add Context to Line Graphs by Adding a Timeline of Key Events

In Chapter 90, I share how to make a timeline in Tableau even when events/marks overlap. One of my favorite applications of a timeline chart is to show my audience when key events occurred in the business that might be affecting performance. As an analyst, I often see large peaks and valleys on trend lines, but they're rarely accompanied by direct context on what might have caused the changes.

With the ability to make worksheet backgrounds transparent, we can now add a timeline directly to a line graph or area chart to provide that meaningful context. Again, this provides a professional design while also serving a practical purpose.

Suppose that we have added this trend to our dashboard:

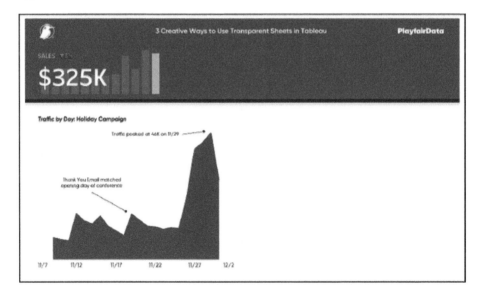

By adding annotations, this trend has already been improved. We see when traffic peaked, which is an alternative way to display the y-axis range, and we also see one bit of marketing context. Even though this chart is helpful, let's make it even better by adding a timeline of key events on top of the trend. After all, this chart represents our critical Tableau Conference and Black Friday marketing push, so it makes sense to let our stakeholders know which marketing tactics were in play during this period.

Here's how the chart looks after we add the jittered timeline view on top of the trend:

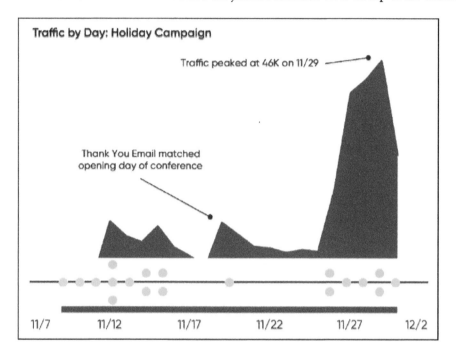

With the timeline layered on top of the trend, our audience can see when key marketing events occurred that might be causing the spikes. Now, we can make this more seamless by changing the background of the timeline to None:

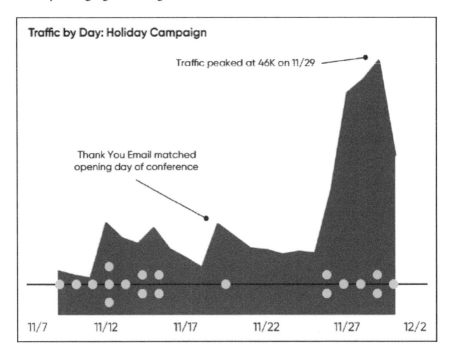

From here, you could color and/or shape the key events by their category and show a guide. But the audience can also get detail about the marketing campaign via the tooltip when hovering over individual marks. The key is they now have context about what's driving performance directly in line with the traditional chart!

CHAPTER 94

3 Innovative Ways to Use Transparent Sheets: Tip 3

Add Context to Scatter Plots by Coloring Quadrants

One of the reasons I like to use scatter plots is because they create a natural, four-quadrant segmentation. If you add a reference line on both the y-axis and x-axis, each mark will fall into one of four boxes. This becomes very actionable because each segment has different attributes (e.g., high sales/low profit, high profit/low sales, etc.) that you can treat differently.

Chapter 23 shows you that you can write a calculated field to color the marks based on which of the four segments they're in—but what if you want to use the mark color to represent something else? You can change the color above and below a reference line, but you cannot use four different colors.

Now, with transparent sheets, we can. This technique works best with fixed size worksheets that have already been added to a dashboard.

First, take a screenshot of the chart and copy and paste the image into PowerPoint or Photoshop:

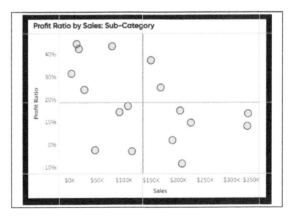

Next, draw a colored square around each area that you want represented as a segment. Add transparency to the colors so that you can see how they will look on the scatter plot. You can add transparency in Photoshop by clicking the layer and reducing the opacity of the layer. In PowerPoint, click the shape and then navigate to Format > Shape Fill > More Fill Colors > Transparency.

Here's how the image looks after we change the opacity to 15% (or 85% transparency) in Photoshop:

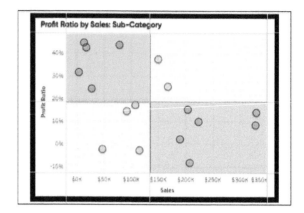

Now remove the original chart from the view in Photoshop or PowerPoint and save the image. If you save the image as a PNG file with a transparent background, you can use this image in Tableau as the top layer or bottom layer:

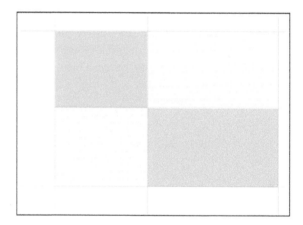

After you save the image, add it as an Image object to your Tableau dashboard and float it over the scatter plot with its original dimensions. If you saved the image as a PNG file with a transparent back-ground, you will be able to see the underlying scatter plot:

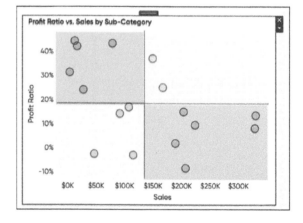

This works visually, but because the image is floating over the scatter plot, you can-not hover over the data points to read their tooltips. If you want to be able to use the tooltips, there are two last steps. First, move the Image object to the bottom layer by selecting the object and then, in the upper-right corner, click the down arrow, hover over Floating Order, and then choose Send to Back:

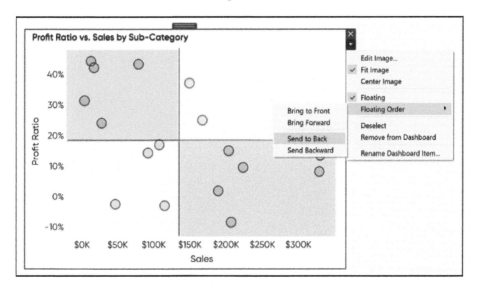

Lastly, as with the first two applications of transparent sheets shared in previous chapters, make the scatter plot worksheet transparent so that you can see the image file that is now below. Now in the final view, we can see my four different segments and hover over marks to read their tooltips:

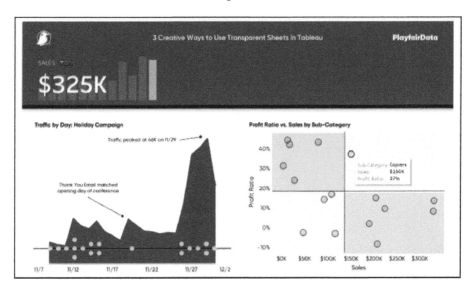

If you prefer to make the image a permanent part of the scatter plot, see Chapter 42.

CHAPTER 95

3 Innovative Ways to Use Set Actions: Tip 1

At the end of Part III, we covered several innovative ways to use parameter actions. *Set actions* work similarly, but instead of overwriting predefined user-generated values (i.e., parameters), they overwrite values in sets. They also have the added benefit of allowing your users to do a multiselect, leading to some even more flexible experiences. The next three chapters show you three innovative ways to use the Change Set Values dashboard action.

You've likely heard of *sheet swapping* in Tableau; a tactic that allows your users to choose what type of chart is shown on a view (Chapter 46). With the introduction of set actions in Tableau 2018.3, you can now also do what I'm calling *button swapping*.

Suppose that you are using images to filter marks on a view. You can now improve the user experience by changing the image for the selection so that it appears to be highlighted!

How to Do Button Swapping in Tableau

By the end of this chapter, you will be able to re-create the user experience from my visualization, *BLOCKBUSTER* (*https://oreil.ly/I_k8W*). By default, each button's background color is gray. When a button is clicked to filter the view, the background color is swapped for red to show what has been selected.

Related: *Practical Tableau*, Chapter 47, "How to Create Icon-Based Navigation or Filters" (O'Reilly, 2018)

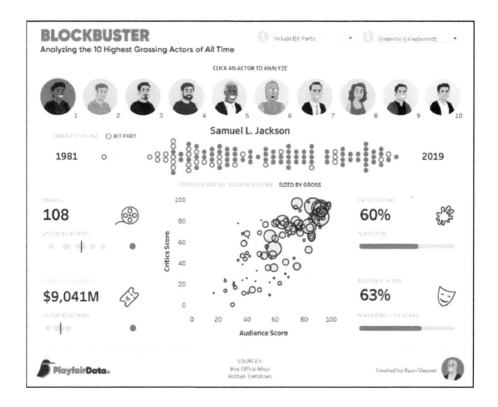

The foundation of button swapping involves using images to filter marks on a view, so it would be helpful to know how to create icon-based navigation or filters in Tableau as a prerequisite to improving on this user experience.

For this tutorial, let's start a new workbook and re-create 5 of the 10 actor buttons from *BLOCKBUSTER*. You can follow along using any images you want.

You need two sets of images to act as your buttons. The first set of images is for buttons in their unselected state; the second set is to highlight selections. Here's how my two sets look for my five actor buttons/filters:

Next, save or copy and paste these images into the following directory on your computer: *Documents\My Tableau Repository\Shapes*. Each folder within *Shapes* is a custom shape palette that you can use in Tableau Desktop. You can save the images into an existing custom palette, but you can also create a new folder and name it something intuitive so that you can find it later in Tableau Desktop. Here's how my custom shapes palette looks for *Button Swapping Example*:

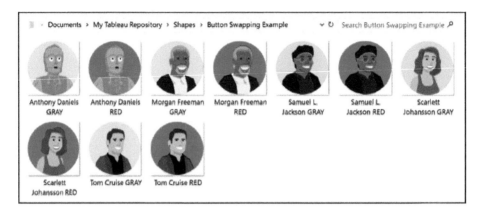

Now we're ready to move to Tableau Desktop and create the navigation. We need to create a set that will eventually be used to inform Tableau what has been selected. Let's create this example from a dimension called Actor: in the Dimensions area in the Data pane, right-click the Actor dimension and then, on the menu that opens, hover over Create, and then click Set. For now, I'm not going to add anything to the set because I want to map shapes for each button in its deselected state:

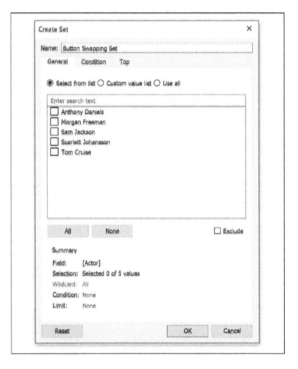

Sets are Boolean, meaning that there are only two outcomes: the dimension member is either in the set or not. Behind the scenes, if the dimension member is in the set, it's classified with a *t* that represents True. When it is out of the set, it's assigned an *f* for False.

Here's the trick. Create a calculated field that equals the string of the set we just created (which will generate a *t* or *f*), plus the dimension that's being used for the buttons. This calculation creates a dimension member for each button preceded by a *t* or *f* to classify what has been selected:

STR([Button Swapping Set]) + [Actor]

Next, set up a control sheet that includes the calculated dimension members that will be used as your buttons. The most important aspect of this control sheet is that the mark type needs to be set to Shape so that you can map each dimension member to a specific custom shape.

Place the newly created Button Swapping Dim field on the Columns Shelf to separate our marks, change the mark type to Shape, and add the Button Swapping Dim field on the Shape Marks Card so that each dimension member is assigned a unique shape:

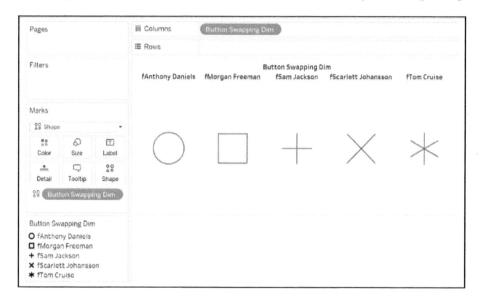

Notice how every dimension member is preceded by an *f*. That's because nothing is selected in the set. These are our buttons in their unselected state.

To map your custom images to each dimension member, click the Shape Marks Card. Click the Select Shape Palette drop-down box, and you will see your custom shape palette that you added to your Tableau Repository:

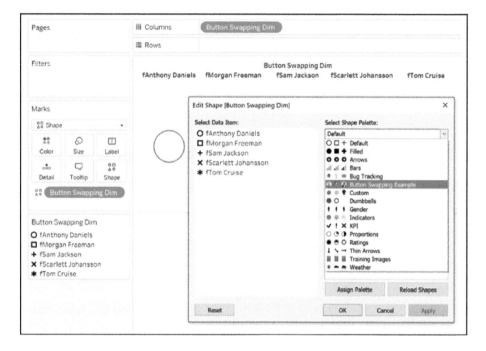

If you already had Tableau open while you were adding a folder to your Tableau Repository, you will need to click the Reload Shapes button in order to see your new palette in the list.

To map the images, simply click each dimension member and then click the image to which it should be mapped. We haven't selected anything yet, so this first set should be mapped to your images in their unselected state:

At this point, our navigation is in its unselected state:

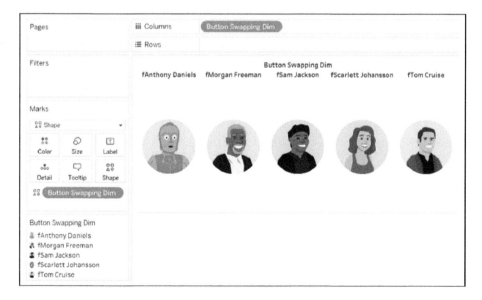

To map each dimension member to its selected state, edit the set so that it includes every dimension member:

Notice that each dimension member is now preceded by a *t* because it is *True* that it is in the set. We also lost the image mapping from the previous step because those images were mapped to the dimension members preceded by *f*. Let's map this set to my images with the red backgrounds to highlight that they have been selected:

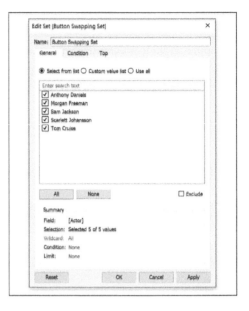

To get this to work with set actions in the next step, we need to ensure that all of the fields being used in the set action are on this sheet. My set is based on the Actor dimension, so I will add that to the Detail Marks Card:

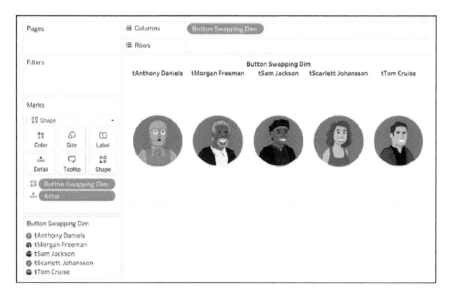

We're now ready to use this as a dashboard navigation. First, add the sheet to a dashboard:

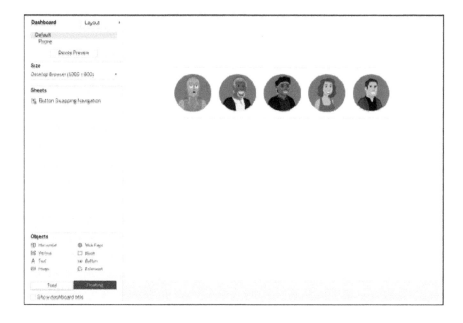

Then add a set dashboard action by navigating to Dashboard > Actions > Add Action > Change Set Values. I like to code my set action so that it leaves the values in the set. This way, the selection persists if you click somewhere else on the view:

Now, if I click a specific button, that dimension member overrides the selections in the underlying set.

So in this case, clicking one actor will make it part of the set, displaying the image with the red background, whereas everything else has a gray background:

By default, these buttons are sorted in ascending data source order. Because the *f* letters that precede all the dimension members that have not been selected are listed first, the highlighted button will always appear last. That's not the best user experience, because the buttons will constantly be switching places.

To permanently set the sort order, go back to the sheet with your images, right-click the dimension member on the Columns Shelf (or Rows Shelf), and then click Sort:

Now the buttons will stay in place and clicking an image on the dashboard will swap out the default image with the gray background for the highlighted red background.

You can incorporate the Button Swapping Set on other sheets so that it acts as a filter when a selection is made and/or add a filter dashboard action so that these buttons control the marks being shown on the dashboard!

CHAPTER 96

3 Innovative Ways to Use Set Actions: Tip 2

How to Drill into a Single Row

Have you ever wanted to drill into a single row of a text table in Tableau? Tableau has nice hierarchy options that allow you to drill down and back up to get varying levels of detail for whatever you're analyzing, but the drilldowns are all or nothing. So by default, if you wanted to click on a Category dimension member to reveal each category's respective sub-categories, *all* the categories will display their sub-categories.

This default behavior becomes problematic with text tables because the number of rows becomes inflated with data that's irrelevant to your analysis. This chapter shows you the three-step solution for drilling into a single dimension member to reveal underlying detail in Tableau.

The trick to clicking into a single row to expand to more detail involves either *set actions* (available as of Desktop 2018.3) or *parameter actions* (available as of Desktop 2019.2). Set actions are more flexible than parameter actions because:

- They allow you to do a multiselect (i.e., drill into one *or* more rows)
- You can use empty sets (parameters always have a current value selected)
- You can instruct Tableau what to do when a selection is cleared

For these reasons, let's set up our drilldown using sets. Let's pretend we're using this table that looks at Sales, Profit Ratio, and Discount by Category from the Sample – Superstore dataset.

We want to drill into individual categories to reveal the Sub-Category dimension for only the selected category.

	Sales	Profit Ratio	Discount
Furniture	$742,000	2%	17%
Office Supplies	$719,047	17%	16%
Technology	$836,154	17%	13%

The first step is to create a set from the dimension you will be clicking to reveal more detail. To create a set, in the Dimensions area of the Data pane, right-click the dimension (Category in this case), hover over Create, and then, on the menu that opens, choose Set. Leave the set blank for now; we will be populating the set during the third step.

Next, create a calculated field that will display the second level of the hierarchy only when the first level of the hierarchy is in the set.

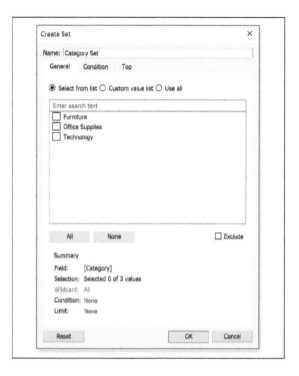

For this example, I want to display the Sub-Category dimension (i.e., the second level of the hierarchy) when a dimension member from the Category dimension is in the set. Here's the most elegant way I can think to write this formula:

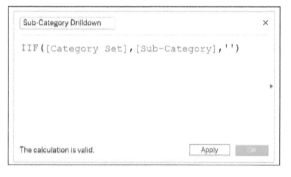

```
IIF([Category Set],
[Sub-Category],'')
```

The IIF function looks at the logic before the first comma; when the statement is true, it displays the outcome between the first and second comma; when the statement is false, it displays the outcome after the second comma. Because sets are Boolean, the Category Set from step one alone is a logical statement; it's essentially a more elegant way to say "IF a Category dimension member matches the dimension member in the set THEN show the Sub-Category dimension ELSE show a blank END."

Let's now place this newly created calculated field onto the Rows Shelf of the table:

Nothing changed because our Category Set is currently empty, and the Sub-Category dimension will be displayed only if a dimension member from the Category dimension matches a dimension member in the set we created in the first step.

In the third and final step, we use a set dashboard action to populate the Category Set, which will create the single-row drill-down effect. To create a dashboard action, the sheet must be on a dashboard, so throw the Text Table sheet containing the Category and Sub-Category drilldown dimensions on a new dashboard:

To create a set action:

1. On the menu bar at the top of the window, click Dashboard.

2. From the list of options, click Actions.

3. In the Actions dialog box, click the Add Actions button and choose Change Set Values.

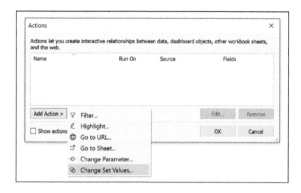

The most important aspect of the Change Set Values dashboard action settings are to specify to Tableau which set you are targeting, but you can also set up when you want the action to execute and what you want to happen when the selection is cleared.

For this use case, I recommend having the action run on Select and "Removing all values from set" when the selection is cleared. This means that the single row drilldown will work when a user clicks a dimension member, and the table will go back as if nothing happened when the selection is cleared:

Now, if a user clicks a specific dimension member, such as the Office Supplies category, the table will drill into the sub-categories for that single row:

		Sales	Profit Ratio	Discount
Furniture		$742,000	2%	17%
Office Supplies	Appliances	$107,532	17%	17%
	Art	$27,119	24%	7%
	Binders	$203,413	15%	37%
	Envelopes	$16,476	42%	8%
	Fasteners	$3,024	31%	8%
	Labels	$12,486	44%	7%
	Paper	$78,479	43%	7%
	Storage	$223,844	10%	7%
	Supplies	$46,674	-3%	8%
Technology		$836,154	17%	13%

When they clear the selection by clicking somewhere else on the view or by pressing the Escape key, the table returns to its default state with no sub-category detail:

	Sales	Profit Ratio	Discount
Furniture	$742,000	2%	17%
Office Supplies	$719,047	17%	16%
Technology	$836,154	17%	13%

In the case that you want to drill into individual rows at the same time, this approach also works with a multiselect by holding the Control key while you click on multiple dimension members. By the way, this works any time the Category, or whatever dimension you are analyzing, is on the view—so it will also work on visualizations.

CHAPTER 97

3 Innovative Ways to Use Set Actions: Tip 3

How to Create Persistent Comparison Sets

Comparisons help us find and communicate insights in Tableau, and segmentation (a type of comparison) is one of the best ways to make those insights actionable. So why not combine these tactics by allowing dashboard users to create custom comparison sets that are relevant to them? This is a user experience that was improved dramatically with the introduction of set actions which allow you to create sets by simply interacting with dimension members on a dashboard. These custom sets persist across dashboard views, providing an opportunity to make use of the same sets across multiple analyses.

This chapter shows you how to design set actions to create persistent comparison sets, but also how to use those comparison sets to provide a visual cue to your audience that reminds them what they have selected and the calculations required to compare a focus dimension member to the comparison set.

How to Create Segments on the Fly with Set Actions

By the end of this chapter, you will be able to create the comparison on the left side of this dashboard, on the fly, by selecting dimension members from the scatter plot:

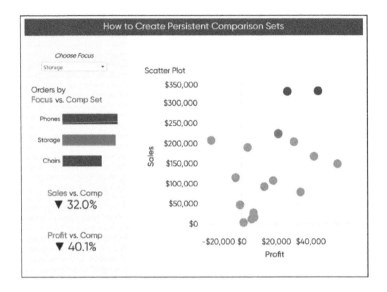

To get started, let's make the scatter plot view that serves as the foundation of the first dashboard. Scatter plots are one of my favorite actionable dashboard elements, and you can learn more about making them more engaging and effective in Chapters 22 through 24. This version is created with the Profit measure on the Columns Shelf, Sales measure on the Rows Shelf, and Sub-Category dimension on the Detail Marks Card:

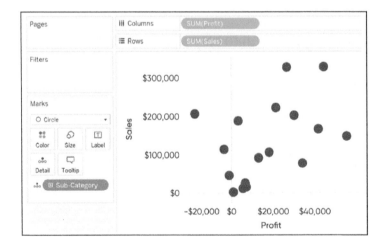

Next, we create a focus dimension member for the workbook. The performance of the focus dimension member will eventually be compared to the comparison set created later in the chapter. My go-to technique is to use parameters to highlight a dimension member on visualizations (see Chapter 40).

Because our focus dimension member will come from the dimension members within the Sub-Category dimension, we can use a shortcut to populate the list of allowable focus values. To do so, right-click the dimension of interest (Sub-Category, in this case), and then, on the menu that opens, hover over Create, and then choose Parameter:

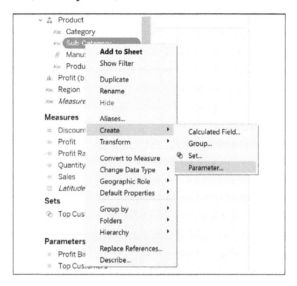

This automatically creates a parameter with a list of allowable values matching the dimension members of the dimension:

To use this parameter to highlight the focus dimension
member (i.e., whichever
allowable value is selected in
the parameter), create a
Boolean calculated field, as
follows. For our example, the
entire formula is:

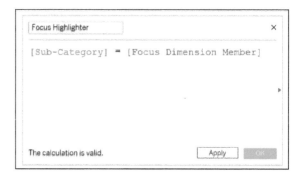

```
[Sub-Category]
= [Focus Dimension Member]
```

If you want to see how this
highlight effect works on its
own, place this calculated field
on the Color Marks Card of
every chart in which you want
to highlight the selected
dimension member. The
dimension member matching
the current value of the
parameter will get one color;
every other dimension member will get a second color:

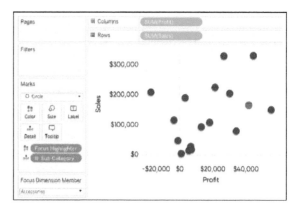

For this illustration, I have chosen to show the parameter control that provides a
drop-down menu with the list of allowable parameter values. You can show the
parameter control by right-clicking the parameter and then choosing Show Parameter Control. You also have the option to use parameter actions to control the selection
of the focus dimension member (see Chapter 2).

Now, we're ready for the comparison set portion of this technique. First, we need to create a set from the dimension being used on the visualization(s). Like our parameter earlier, the easiest way to do this is to right-click the dimension, hover over Create, and choose Set:

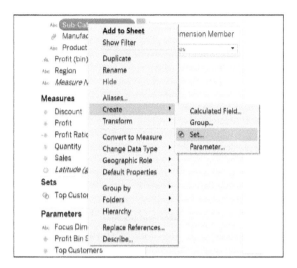

Leave the set empty for now; we will overwrite the values being included in the set with set actions in a future step:

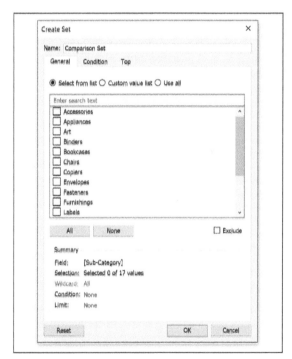

Next comes the key calculated field that drives all our comparisons and future calculations. Here's the formula:

```
IF [Sub-Category] =
[Focus Dimension Member]
THEN "Focus"
ELSEIF [Comparison Set]
THEN "Comparison"
END
```

This calculated field classifies the dimension member matching the current value of the parameter as the Focus and the dimension members in the Comparison Set as Comparison. Note the first line of this code is the parameter highlighter we used to color the scatter plot so far; we're just adding in the logic to classify the dimension members in our set as the comparisons.

Next, let's replace the Focus Highlighter field on the Color Marks Card with the newly created Focus / Comparison calculated field so that the dimension members are classified as one of three things: Focus, Comparison, Null—or not a Focus or Comparison dimension member (i.e., everything else).

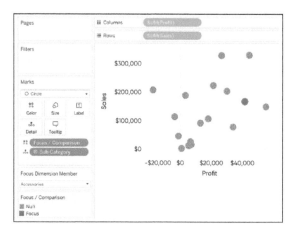

Note that we don't see any dimension members classified as comparisons yet because our set is empty. If we were to add the first five dimension members to the Comparison Set, we would see the Focus dimension member of Accessories (selected via the parameter control) colored red, the four selected dimension members other than Accessories from the Comparison Set colored blue, and everything else colored gray. Again, let's overwrite these selections later using set actions:

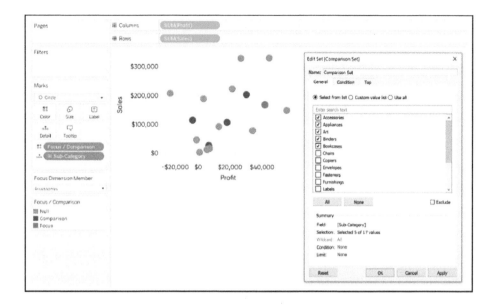

In this example, Accessories was classified as Focus and not Comparison because IF/ THEN logic is processed in order and each dimension member can be classified as only one thing at a time. In our Focus / Comparison calculated field, the Accessories dimension member was classified as Focus before even moving on to the next line of logic.

This next step is optional, but I like to create a visualization that illustrates which dimension member is being used as the focus and which are included in the comparison set. This element is also meant to illustrate the scope of the selections (i.e., has the dimension member selected had a larger or smaller impact on the business).

This can be any visualization, but we'll go with a simple bar chart that looks at Unique Orders per Sub-Category. I've also filtered the marks with the Focus / Comparison calculated field from earlier so that we're keeping only dimension members that have been classified as either the Focus or a Comparison:

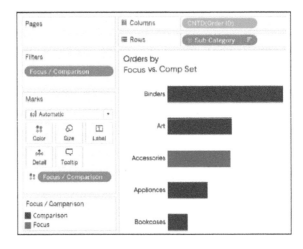

If you plan to create large sets that result in too many bars, you can use the INDEX() function to limit the number of marks included (see Chapter 58) along with your Focus dimension member. If this is of interest, in addition to the Focus / Comparison field on the Filters Shelf, you would need to add a calculated field that looks something like this:

```
[Focus Dimension Member] = ATTR([Sub-Category]) OR INDEX()<=5
```

This would keep the Focus dimension member, plus the first five rows of the bar chart.

Next, let's put this into action by adding both the scatter plot and bar chart to a dashboard:

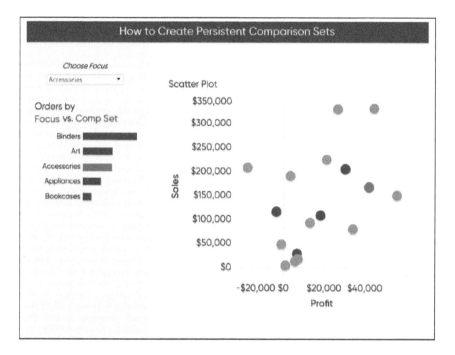

To allow our audience to update the Comparison Set on the fly by simply selecting dimension members on the scatter plot, we use set actions. To create a set action, on the menu bar at the top of the Dashboard interface, click Dashboard > Actions. From within the Actions dialog box that opens, click the Add Action button and then choose Change Set Values.

While designing a set dashboard action, you can choose where the action originates, when you want the action to execute, the set that will be overwritten with the selections, and what you want to happen when the selection is cleared out. In this example, I'd like to click the scatter plot and have the selections overwrite the Comparison Set. When the selection is cleared, I would like the selections to stick until another selection is made. This means that the selection will persist, even when navigating to other dashboards:

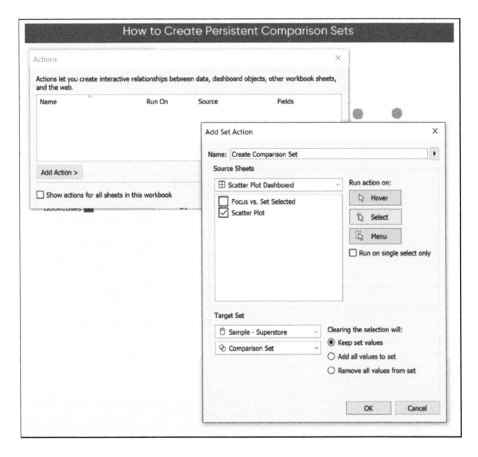

After you click OK to save the new dashboard action, simply clicking the scatter plot will update the sub-categories included in the set, which in turn updates the bar chart. Perhaps instead of adding the first five sub-categories to the set, you would like to create a better peer group by selecting the four dimension members closest to Accessories on the scatter plot.

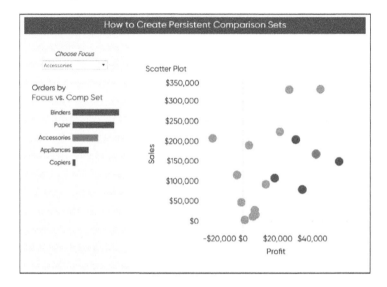

Just think of the possibilities with this. Your audience can now choose a focus of the dashboard that is relevant to them and then create a comparison set based on a peer group, a single benchmark, or a series of benchmarks—all on the fly!

One way I like to use this is to compare the focus to the comparison set with callout numbers (see Chapter 39).

To illustrate, lets create a calculated field that computes the percent difference between the Focus dimension member and the Comparison Set for the Sales and Profit measures. The first calculation would be as follows:

```
(SUM(
IF [Focus / Comparison] = "Focus"
THEN [Sales] END)/
(SUM(IF [Focus / Comparison] = "Comparison" THEN [Sales] END)/
COUNTD(IF [Focus / Comparison] = "Comparison" THEN [Sub-Category] END)))-1
```

The numerator for this calculation isolates the SUM of Sales for the Focus dimension member. The denominator does the same thing for the dimension members classified as Comparison, but then divides that amount by the number of dimension members in the Comparison Set, creating a weighted average. The weighted average would not be required if we were using ratios (i.e., Profit Ratio instead of Profit). The -1 computes the percent difference when the format of this measure is converted to percentages.

Now you can make a callout number in the format of your choosing. I like to use conditional up and down triangles, as outlined in Chapter 48:

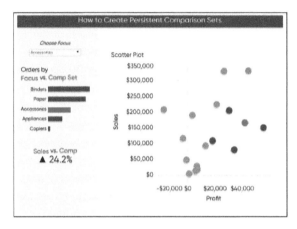

Let's do the same steps for the Profit measure:

Here's where the magic happens. Now the user can select a peer group, a single benchmark, or a benchmark group. Those selections overwrite the Comparison Set. That Comparison Set feeds the bar chart illustrating the scope of the selection and the callout numbers computing the performance of the Focus versus the comparison(s).

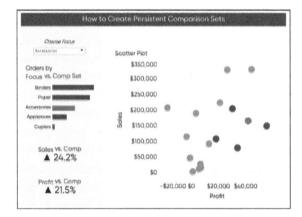

Here's how the dashboard looks after selecting Storage as the focus and choosing the two best performing sub-categories as the benchmark to see what it would take to make Storage rise in the rankings:

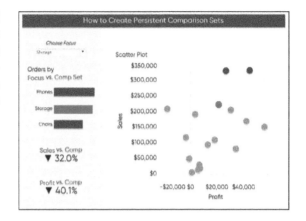

This is becoming a very rich analysis. We can now see visually where Storage is compared to the selected benchmarks on the scatter plot. We also see that, of the three selected dimension members, Storage ranks second in orders indicating that this is an important sub-category to the business. Unfortunately, Storage is lagging in Sales and Profit compared to the benchmark set by 32% and 40.1%, respectively.

Lastly, this functionality persists across dashboard views, so you can just duplicate the first dashboard, leave the left side as is so that your audience is oriented with the segmentation being included, and then add alternative visualizations:

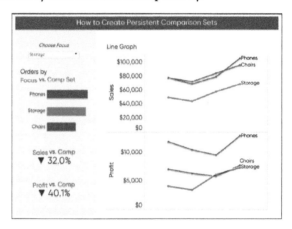

CHAPTER 98

3 Innovative Ways to Integrate Other Products: Tip 1

The final frontier for innovating in Tableau is to integrate third-party products that complement the strengths of Tableau Desktop. Chapter 10 explains how Tableau has opened up the software so that third-party developers can extend the default dashboard functionality, but there are several opportunities to integrate other products without developing or using custom-built extensions.

These integrations allow us to take Tableau to another level, making your dashboards stand out from both a design and user experience perspective. The next three chapters will discuss ideas for using non-Tableau products for enhancing your Tableau dashboards. You will see how to use design elements made with Photoshop or Power-Point, combine dashboard actions with Google searches, and create a dynamic commenting system using Google Forms.

How to Complement Dashboard Design with Third-Party Design Elements

Chapter 4 introduces the concept of using an icon as a Button object; for example:

Chapter 89 takes the concept a step further and integrates icons as mark types to create dynamic indicators:

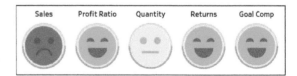

Both tactics are accomplished with royalty-free images purchased from The Noun Project (*http://thenounproject.com*) and iStock Photo (*http://istockphoto.com*), respectively. You can also use images from outside of Tableau to take Tableau's default design options to a new level.

Because Tableau dashboards allow you to add Image objects and sheets can be set to float and be transparent, we can enhance dashboard elements with images built in Photoshop or PowerPoint. As one simple example, consider this series of cards that you might find along the top of a dashboard:

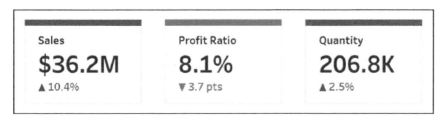

By default, Tableau sheets and layout containers have formatting options including border colors, background colors, and padding. But let's say that you want to add in rounded corners and an outer glow effect around the cards. You could make this image in Photoshop:

If you save this image as a PNG file type with transparency, the checkered area will be empty. This means that if you float the image over the three cards as an Image object, you'll have the custom border effect, but still be able to see the underlying sheets.

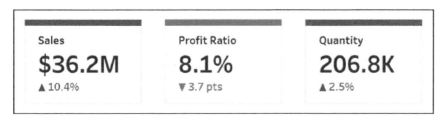

We've just created custom designed callout cards by simply floating an Image object over the existing sheets. Also, you can set a destination URL to each Image object, meaning that you can use these custom designs as buttons to link to deeper analysis on Tableau Server!

Even if you don't have the ability to create transparent PNG image files in Photoshop, you can create the images in PowerPoint, take a screenshot of the images, and save them as JPG image files. You then can add these files to a dashboard as Image objects, and they would become the background of the existing sheets. You would float the sheets over the images and set their backgrounds to transparent as described in Chapter 5.

The latter approach has its own advantages, including the ability to see tooltips on existing sheets and the option to set dashboard actions to run when the sheet is clicked.

CHAPTER 99

3 Innovative Ways to Integrate Other Products: Tip 2

3 Ways to Use Google with Dashboard Actions

Google has a number of powerful tools that can help with your business analyses. In addition to the native Google Analytics Tableau connection with which you can create customized Tableau dashboards from your web analytics data, other Google services can be integrated with Tableau using dashboard actions. This chapter shows you how to do an image search, view your paid search ads, and lookup a Google Trend, all from within a Tableau dashboard.

The following tips have several benefits, including:

- Efficiently adding a visual impact to your dashboards
- Providing context to you and your end users
- Improving your analyses with very little manual effort

How to Add a URL Dashboard Action to a Tableau Dashboard

To do a Google image search, view your Google paid search ads, or lookup a Google Trend, you will need to add a URL action to your Tableau dashboard. To lay the foundation for this tutorial, I am going to make a simple dashboard.

First, let's place a horizontal layout container on a new dashboard. Within that layout container, place a bar chart showing Sales by Sub-Category and a Web Page object:

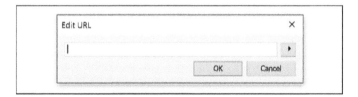

Note that when you add the Web Page object, Tableau asks you to enter a URL, but for now you need to leave this blank.

At this point, our dashboard looks like this. Note the blank space on the right side of the horizontal layout container is the Web Page object. This will soon be filled with a Google image search, Google paid search ads, or Google Trends:

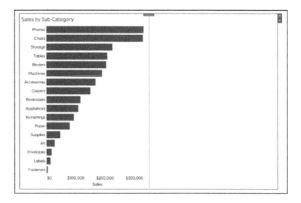

To add a URL dashboard action, on the menu bar at the top of the window, click Dashboard > Actions and then, in the Actions dialog box, click Add Action, and then choose Go to URL:

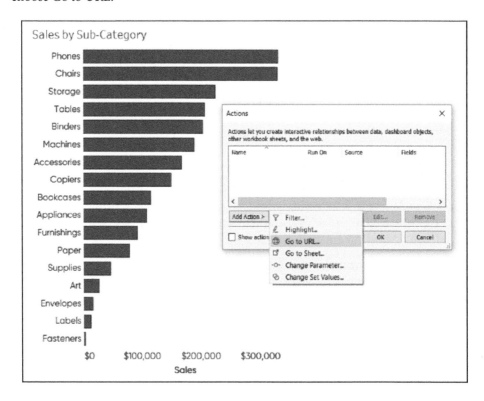

Ways to Integrate Google with Tableau URL Actions

When doing a Google image search, a Google search (that contains paid search ads), or a search for a Google Trend, a URL is generated with a query string. Here is how each respective URL looks when searching for the top Sub-Category (Phones) from our dashboard:

Google Image Search: *https://www.google.com/search?q=phones&tbm=isch*

Google Search: *https://www.google.com/search?q=phones*

Google Trends: *https://trends.google.com/trends/explore?q=phones*

I'm consolidating all three examples because they all work the same way. Depending on which of the three tactics you want to execute, you will use the aforementioned corresponding URL. Here's how our URL dashboard action looks if we want to set up a traditional Google search to see whether our paid search ads are showing up with our competitors' ads:

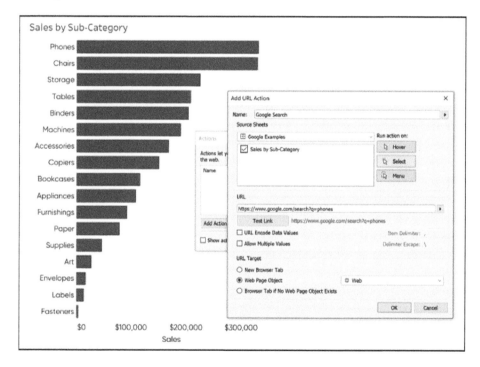

Here's where the magic happens.

Instead of keeping phones as a static search, we can replace the text after "?q=" with a field from our dashboard by clicking the arrow to the right of the URL box:

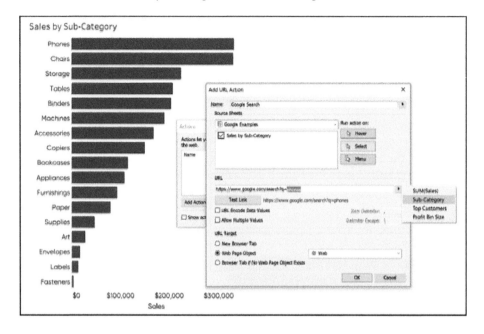

Here's how the URL dashboard action looks after we replace *phones* from the query with the Sub-Category dimension from our dashboard:

The way this is set up, now when you click on any of the bars on the dashboard, the Web Page object on the right will be populated with a Google search! Here's how the dashboard looks after clicking the Tables sub-category:

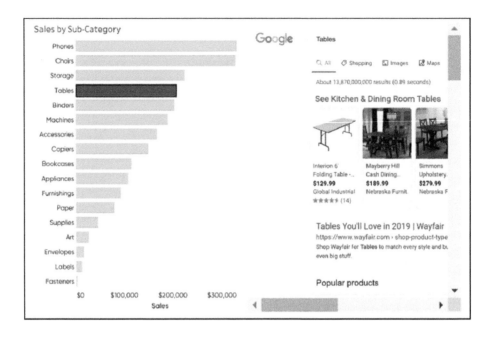

Here's how the same view would look if you used the Google *Image* Search URL instead:

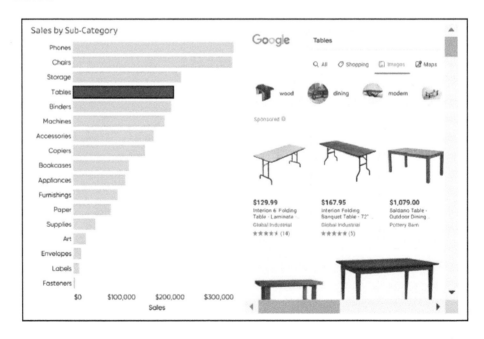

Lastly, here's how the same view would look if you used the Google *Trends* URL:

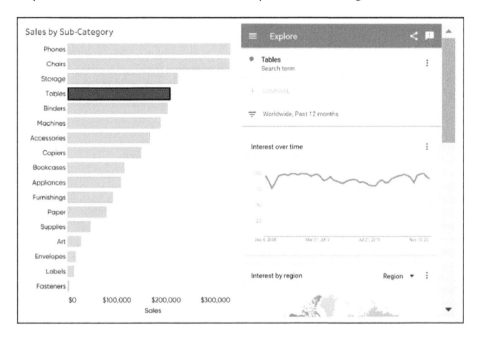

If you don't like the way the web page looks within the Tableau dashboard, follow the same steps from this tutorial but *do not* add a Web Page object. If there is no Web Page object to fill with the URL from the URL dashboard action, a new browser window will open outside of Tableau, instead. Another option is to change the Go to URL dashboard action settings to target a new browser tab instead of the Web Page object.

CHAPTER 100

3 Innovative Ways to Integrate Other Products: Tip 3

How to Make a Commenting System with Google Forms Integration

Chapter 84 shows you how to add custom integrated insights to a Tableau dashboard. This tactic allows you to communicate insights and recommendations directly on dashboards. There are two drawbacks to this approach:

- You must direct the user to a different view to enter their commentary
- The text is static until the comments are updated manually

This chapter shares a new approach to adding custom comments directly to a Tableau dashboard by integrating Google Forms and using layout containers to create a seamless commenting system. On top of that, you can filter comments by date and user!

To illustrate how to create comment cards in Tableau, Let's re-create the right side of this variation to the *Super Sample Superstore* dashboard (*https://oreil.ly/6Unk0*):

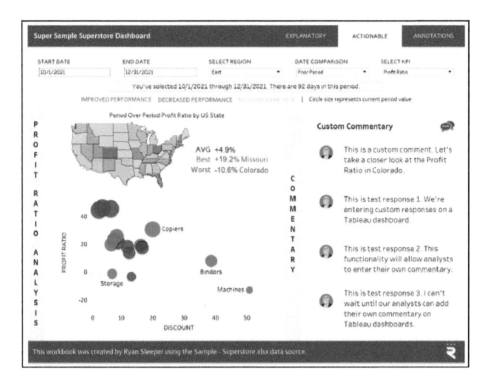

First, set up a Google Form, a free tool that comes with Gmail accounts. To find or add to your Google Forms, in the upper-right corner of your Gmail account, click the Google apps menu next to your avatar and then click the More button:

You can add questions to a Google Form by clicking the plus sign in the upper-right corner of the form. I recommend adding at least one Short Answer question and one Multiple Choice question. The short answer will eventually be the comment displayed on a dashboard, and the multiple-choice question can be used to have the analyst select who entered the comment:

When your form is ready, in the upper-right corner of the browser, click Send, navigate to the link tab, copy and paste the share link into a different browser, and enter two or three test responses. These test responses will allow us to lay the foundation of the comment sheet being used on the dashboard. I also recommend you take note of the share URL because you will need it in a future step.

Next, go back to where you created the Google Form, click the Responses tab, and then, in the upper-right corner of the form, click the Google Sheets icon to convert the responses to a data source:

Back in Tableau Desktop, connect to a new data source and connect to the Google Sheet you created in the preceding step. For best results, leave the connection as Live so that the latest responses can be brought into the dashboard with a quick refresh:

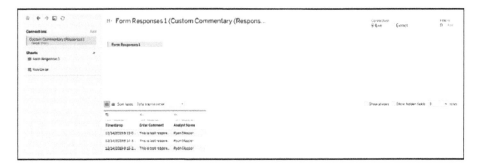

Create a new sheet with a table of responses formatted as you prefer:

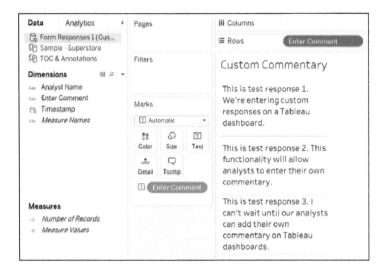

Note that Google Forms automatically creates a dimension called Timestamp with which you can filter the comments to relevant timeframes. You can also use an INDEX() filter (see Chapter 58) to keep only the N most-recent comments. When the worksheet containing the comments is ready, place it on your dashboard:

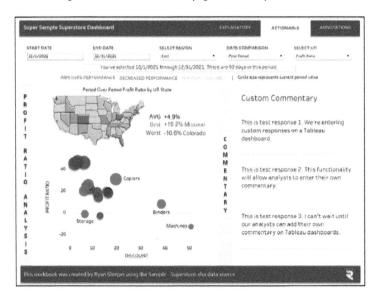

Herein lies the magic.

Place either a horizontal or vertical layout container as a floating object on top of the dashboard wherever you want the analyst to enter their insights. Then, place a Web Page object inside of the layout container and copy and paste the Google Form share link that you received earlier as the URL:

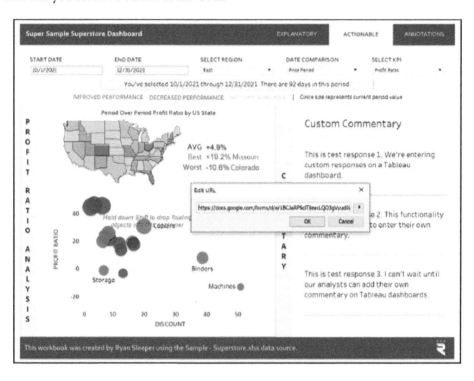

This will frame the Google Form into a Tableau dashboard where you can include additional comments:

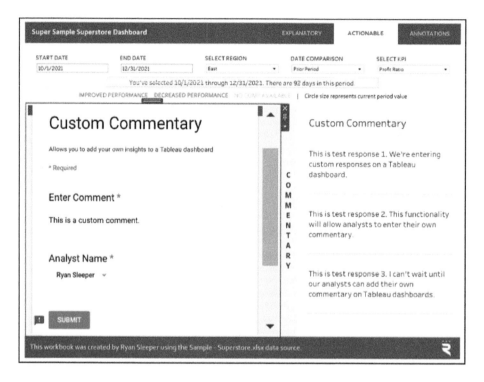

After you refresh the data source by navigating to Data > [the data source] > Refresh, the new comment appears on the dashboard:

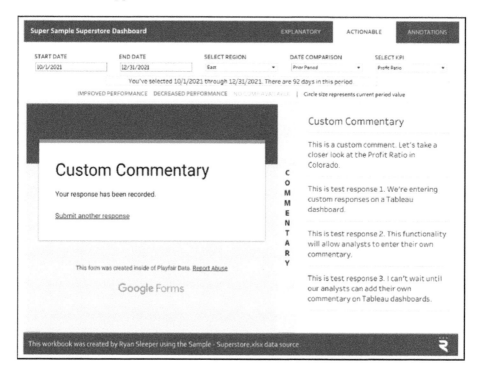

This works even better on Tableau Server or Tableau Online because the user simply needs to click the Refresh button in the top of the browser window to get the latest comments to appear.

Now that all of the pieces are in place and working, I suggest adding a Show/Hide button (see Chapter 6) to the layout container with the Google Form. This will allow you to toggle the custom commentary form on and off so that it does not distract from the view until an analyst is ready to enter a comment.

To add a Show/Hide button to a Tableau layout container, select the layout container and then, in the upper-right corner of the selected container, click the down triangle to access its options, and then choose Add Show/Hide Button:

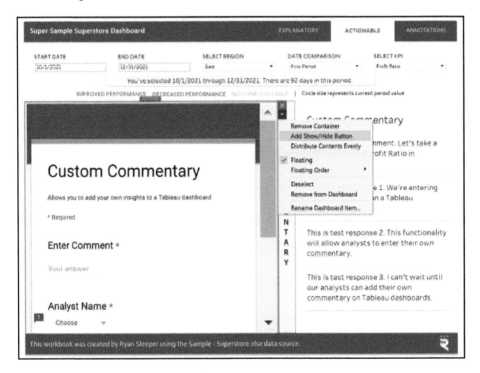

This button can be mapped to different images depending on whether the container is being shown or hidden and the button can also be moved to wherever you want within a dashboard. Following is a screenshot that shows how our view looks after mapping a speech bubble icon to the button and moving it so that it better aligns with the commentary area.

Note the Show/Hide button is a separate object from the container with the Google Form, so I can keep the form where it was with its larger dimensions, but create a better user experience by having the button in line with the Custom Commentary title.

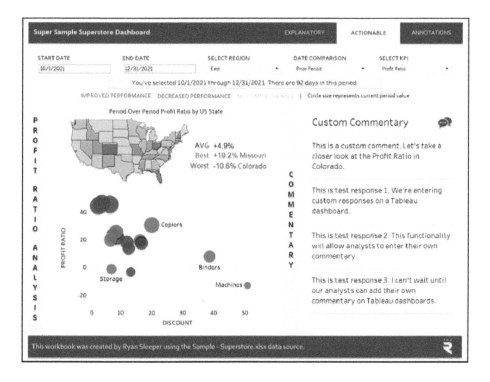

If a user clicks the speech bubble, the layout container will be shown, and thus, the Google Form where they can add comments will appear. I've mapped a red speech bubble for when the layout container is in an open state. The user can close the container and view the underlying visualizations by clicking the Show/Hide button again:

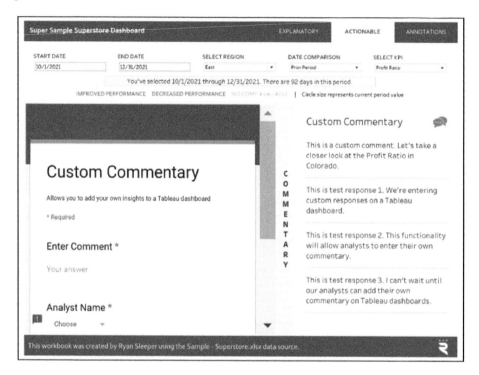

Related: *Practical Tableau*, Chapter 47, "How to Create Icon-Based Navigation or Filters" (O'Reilly, 2018)

Want to innovate with this even more?

Duplicate the sheet containing the comments, change the mark type to Shape, remove the Enter Comment dimension from the Text Marks Card, and Add the Analyst Name field to the Shape Marks Card. You can map avatars to different analyst names so that they appear alongside their comments.

Here's how the final view looks after putting my custom avatars in line with their respective comments on the dashboard, formatting, and closing the layout container with the Google Form:

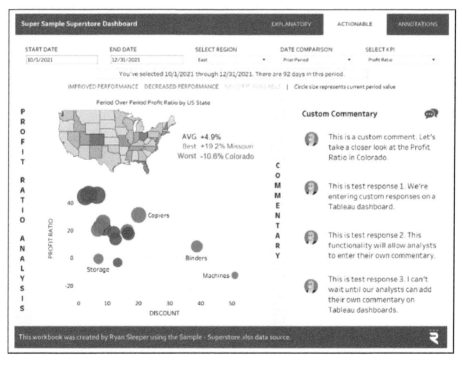

We've just created a dynamic commenting system that allows analysts to add their own commentary directly to a Tableau dashboard in real time! These comments can be filtered by date or analyst to display the most relevant insights and actionable recommendations.

Chapter 201

When I began writing *Practical Tableau*, the name of the project was Tableau 201. The number 201 represented two objectives. First, I was targeting 201-level material, as opposed to 101-level beginner topics or 601-level master topics. I know that by now most analysts have some foundational Tableau knowledge and are primarily looking to level up, but they also have a need to create practical applications that sometimes require more advanced topics to be translated for them so that they are accessible.

Second, I wanted to provide 201 different tutorials. Candidly, there have been times during the past five years when I was skeptical that I would ever reach that number. But Tableau has created such a flexible data visualization and analytics program that I am now convinced a little bit of problem-solving and elbow grease will lead to infinite solutions.

So here we are—at the end of *Innovative Tableau*—and 200 chapters later. I've decided to leave "Chapter 201" for you to write.

How can you innovate from here? What new technique can you share with us to collectively improve our understanding of data? How can you push your organization or cause to new heights by using data visualization and Tableau to translate raw data into something meaningful?

I think Ben Jones, founder of Data Literacy, put it best in the foreword of my first book when he said:

> The world in which we live—our communities and our planet itself—are depending on each one of us becoming not just fluent in the language of data, but eloquent.

Thanks for reading,

—Ryan

Index

sort order
    changing with buttons, 536-542
    spaces versus letters, 289
    top versus bottom N lists, 451
spacing, default versus custom, 376
spark bars, 544-547
splash pages, 433-436
statistical anomaly detection, 465-470
stock ticker gauge, 238-244
Story Points feature, 437, 440
Streets map style, 38

T

table calculations
    reverse-engineering calculations, 363-366
    saving for future use, 370
    updating on the fly, 366-369
table columns
    adding tooltips to dimensions, 336-343
    creating custom headers, 66-69
    increasing number of, 63-65
Tableau
    benefits of, xiii, 363
    new and updated features
        Button (Navigation) dashboard object, 19-24, 428-432
        Change Parameter action, 4-8, 413-416
        Change Set Values actions, 9-18
        dashboard extensions, 52, 582
        Explain Data, 43-47
        mapping capabilities, 38-42
        Show/Hide buttons on layout containers, 31-37
        transparent sheets, 25-30, 543-553
        upcoming features, 53
        Viz in Tooltip, 48-51, 329-335, 446-453, 524-526
    prerequisites to learning, xiv
    tips for learning, 1-3
    versions of, 3, 344, 427
    video tutorials, 1
Tableau community, 2, 19
Tableau Desktop, versus Tableau Public, 3
Tableau Public, 3, 19, 230
Tableau training workshops, 3
tables (see also graphs; charts; maps)
    highlight tables
        color-coded navigation, 502-506

coloring highest and lowest points on, 454-459
colors, based on discrete segmentation, 77-81
colors, using in moderation, 75
creating, 72
formatting cells, 74
text tables
    adding tooltips to dimensions, 336-343
    custom headers, 66-69
    expanding single rows in, 564-568
    formatting like finance reports, 288-290
    increasing number of columns in, 63-65
    text alignment, 70-71
text
    aligning in tables, 70-71
    concatenating text fields, 496
    conditionally formatting, 285
    custom fonts, 283
    custom indentation, 288-290
    performance indicator titles
        creating, 203-206
        with comparison score, 207-213
    removing redundant labels, 61
    removing varying font styles/formats, 59
text tables (crosstabs)
    adding tooltips to dimensions, 336-343
    aligning table text, 70-71
    creating custom table headers, 66-69
    expanding single rows in, 564-568
    formatting like finance reports, 288-290
    increasing number of columns in, 63-65
third-party design elements, 582
tile maps (see trellis maps)
timelines
    best use of, 161
    creating, 161-165
    key events on line graphs, 548
    for overlapping events, 527-535
    reference lines for current date, 165-167
tooltips
    adding call to actions in, 434
    adding images to, 329-335
    adding to dimensions, 336
    displaying visualizations in, 48-51
    drawbacks of, 344
    dynamic, 344-347
    showing top 10 lists in, 446-453
top N filters

About the Author

Ryan Sleeper is founder and principal at Playfair Data and has consulted with dozens of the world's best-known brands. He is author of the book *Practical Tableau: 100 Tips, Tutorials, and Strategies from a Tableau Zen Master* and has shared his data visualization strategies worldwide, including speaking engagements in London, Tokyo, Toronto, and Singapore. Ryan's work in Tableau has earned him the titles of Tableau Zen Master, Tableau Global Iron Viz Champion, and Tableau Public Visualization of the Year author.

Colophon

The birds on the cover of *Innovative Tableau* are green thorntail hummingbirds (*Discosura conversii*), one male and one female. Hummingbirds have over 345 species and are only found in the Western Hemisphere. Most species, like the green thorntail, live south of North America, and 20 are found only in Central America. Green thorntails inhabit humid forests and flowered clearings from Costa Rica to Ecuador.

A typical hummingbird is three to five inches long, the smallest of all birds. Green thorntails are small even among hummingbirds at two to three inches long and three grams in weight. The long, forked tail adds another inch and a half to the male green thorntail. Both sexes have green feathers from the crown of the head to the base of the tail, where a white band separates the green from deep dark blue. Their wings are dark as well. Female green thorntails build nests and incubate eggs independently.

Hummingbirds seek out brightly colored, scented flowers, from which they extract nectar with their tubular, extendable tongues. Green thorntails likely also feed on small insects. Bumblebees and hawk moths share similar diets, and green thorntails compete with them for territory.

Green thorntail hummingbirds have a conservation status of Least Concern. Many of the animals on O'Reilly's covers are endangered; all of them are important to the world.

The color illustration is by Karen Montgomery, based on a black and white engraving from *Wood's Natural History*. The cover fonts are Gilroy Semibold and Guardian Sans. The text font is Adobe Minion Pro; the heading font is Adobe Myriad Condensed; and the code font is Dalton Maag's Ubuntu Mono.